T5-ARJ-314

GARLAND STUDIES ON

INDUSTRIAL PRODUCTIVITY

edited by

STUART BRUCHEY
UNIVERSITY OF MAINE

A GARLAND SERIES

THE TELEGRAPH

How Technology Innovation Caused Social Change

ANNTERESA LUBRANO

GARLAND PUBLISHING, Inc.
New York & London / 1997

Library of Congress Cataloging-in-Publication Data

Lubrano, Annteresa, 1955–
The telegraph : how technology innovation caused social change / Annteresa Lubrano.
p. cm. — (Garland studies on industrial productivity)
Revision of the author's thesis (Ph. D.)—Graduate School and University Center, CUNY
Includes bibliographical references and index.
ISBN 0-8153-3001-4 (alk. paper)
1. Telegraph—History. 2. Telecommunication—Technological innovations. 3. Technological innovations—social aspects. 4. Information technology—Social aspects. I. Title. II. Series.
HE7631.L83 1997
384.1'09—dc21 97-25292

Printed on acid-free, 250-year-life paper
Manufactured in the United States of America

Dedication

This book is dedicated to the late

Sidney H. Aronson

Contents

Tables and Figures

Preface

Information processing is crucial to social life and an important element of control. Innovations in information processing have the potential to dramatically alter social relations. Understanding the process of technology innovation and diffusion as well as the economic, social, political and cultural impact of a diffusing/diffused technology is crucial to understanding society as technology is often the impetus for social change. This book addresses both the process and impact of technology innovation as it relates to communication technology. Part I examines prominent technology innovation models and suggests a new model with a diffusion phase specific to interpersonal communication technology. The telegraph is used as a case study as it was the first modern communication innovation that dramatically altered society's ability to control information processing and in the process altered relations of control within society. The telegraph was a prototypical of subsequent interpersonal technology innovations such as the telephone, fax machine, electronic mail, and the personal computer and Internet systems. Parallels with these technologies are drawn throughout the book. Part II addresses the impact of the telegraph on social institutions such as work, the judicial system, culture, politics, the medical system, social stratification and globalization. The implications of ownership and control raised by telegraphic technology are just as germane to pc and Internet technology today and can be used to illustrate the revolutionary impact of the diffusion of Internet technology. This book is a journey rooted in the past with applications and implications for the present and the future

Introduction

The craving for instantaneous communication extends far back in history. Early historians write about the Greeks perched on rooftops watching for fire signals, or those who used a glint of sunlight on a polished surface to send messages across distance. The Moors in Algeria were using the heliograph, a device for signaling by flashing the sun's rays from a mirror, as far back as the eleventh century A.D. Ancient Romans used citizens perched on towers as a form of communication to bridge the miles. Societies of the 1800's used semaphores to communicate across distance. Why this obsession with instantaneous communication? Having information or knowledge sooner rather than later was to someone's benefit and, to adapt Beniger's (1986) terminology, probably allowed greater societal control. However, no matter how great the desire for instantaneous communication was, it was always limited by distance as well as the vagaries of nature. Time-of-day, as well as weather conditions such as rain, snow or clouds, hampered the ability to transmit information if it depended on visual communication. For nearly all of human history instantaneous communication was beyond human control. The invention of the electric telegraph changed all this. The telegraph gave human agency to the control of information. Beniger (1986,: 8) tells us that "Because both the activities of information processing and communication are inseparable components of the control function, a society's ability to maintain control - at all levels from interpersonal to international - will be directly proportional to the development of its information technologies." Burke (1978,: 81) also indicates that "The economic, political and cultural life of a country is shaped by the way it uses its communications systems to encourage or deny access to information which such systems transmit."

The telegraph was the first modern communication innovation that significantly altered the way people interacted with each other. The telegraph was the first step in a process of continuous improvement which led to the telephone and continues today with the "information superhighway". The telegraph can actually be considered the grandfather of the data superhighway - the telegraph operated on a digital format (on/off mode); the telephone, the next major innovation in communication technology, operated in an analog mode; and the data superhighway is again dependent on digital processing. The data superhighway is one more step in the quest to mediate or annihilate the effects of distance. Because of the importance of new technologies, much sociological, economic, and business research has been conducted on the origin, diffusion, and impact of new technologies. Various aspects of this process have been modeled, implicitly if not explicitly. One purpose of this dissertation is to collect these multidisciplinary models into a more comprehensive sociological framework with the aim of codification and insuring that gaps in sociological understanding are catalogued and at least partially filled. Another purpose of this dissertation is to help identify the socioeconomic factors which retard/speed the adoption of new technologies. Speed is an increasingly important competitive factor and hence the choice of the telegraph as it was the first major technological innovation that affected the speed of communication. The models and concepts to be discussed include: S-curves, push-pull (economic theory), product life cycle and profit cycle models of diffusion, and adoption models of diffusion.

The telegraph is a strategic research site for understanding the generation and impact of technologies because it represents the *first* in a series of innovations which gave man more control over long distance communications (for example, messengers, carrier pigeons, optical telegraphs) . Additionally, another aspect of its typicality lies in issues of control that are generated by new communication technologies. The telegraph, like each communication innovation that arose subsequently, altered relations of control and power. Various researchers have written about the relationship between technology and society. The work of Noble (1977), Beniger (1986), Giddens (1991, 1983), and Fischer (1992) are referenced for the multiple and at times conflicting insights they offer into this relationship. The various perspectives each have something to contribute to our total understanding of the relationship between technology and society. This work will draw on several specialties within sociology, such as, stratification, formal organizations, politics and government.

Noble (1977) tells us there is nothing "natural" or autonomous in the history of technology. Rather, technological change is a social process, a human enterprise and as such technology will be channeled to meet the needs of the dominant institutions in society. In the United States economic institutions are dominant and therefore, Noble suggests, the interests and needs of economic institutions will have stamped technology, defining certain institutional arrangements, uses and access, thus blocking other possibilities. Implicit in Noble's work is the idea that modernity wasn't an accident. The particular character of our modern culture was a conscious product; it is "America by design" not America by invention. This consciousness of agency was intimately tied to control, and as I will demonstrate, communication is essential to control. This work will examine how this particular form of communication technology mediated the relationships among the government, private enterprise, and society and how the struggles for power and control shaped modern society.

Beniger (1986) examines this issue of control. He too believes that modern society is not the result of chance or whimsy. Beniger (1986: 7-8) uses the word control to refer to "purposive influence toward a predetermined goal" or "any purposive influence or behavior, no matter how slight." He indicates that information processing and reciprocal communication are inseparable from the concept of control. Beniger says that information processing is inherently goal-directed, and hence, essential to purposive activity. Additionally, he states that two-way interaction between controller and controlled is also critical. Consequently, Beniger (1986: 8) states: "So central is communication to the process of control that the two have become the 'joint subject of the modern science of cybernetics . . . -' the entire field of control and communication theory." Beniger looks at a very fundamental level of technology - information processing - and concludes that it has resulted in what he labels the "Control Revolution". The changes brought by the industrial revolution and the increasing complexity of capitalist, material economy led to a crisis of control and the need for integration within society. The development of new communication technologies led to a change in the nature of control in society from one based locally to one characterized by centralization. The telegraph was the first modern communication technology with this centralized character which was able to alter the relations of power and control in society. (It made long distance control more viable and direct.) Beniger's work suggests that in examining the impact of a new communication technology one should be concerned with the question of how it increases control. This idea will inform my look at the history of the telegraph in America.

Anthony Giddens (1983) posits that power is an inherent component of interactions and that power is generated in and through the reproduction of structures of domination, which includes the dominion of human beings over the material world (allocative resources) and over the social world (authoritative resources). Giddens (1983: 91-92) Power is generated by the transformation/mediation relations inherent in the allocative and authoritative resources comprised in structures of domination. Storage of authoritative resources involves the retention and control of information or knowledge.

Giddens also says that structures can be analyzed in terms of the transformations and mediations in human activity through which they are in turn sustained.

> "Transformation and mediation: the two most essential characteristics of human social life. Transformation capacity ... forms the basis of human action and at the same time connects action to domination and power. Mediation expresses the variety of ways in which, in social systems, interaction is made possible across space-time. All interaction is carried across time and space by media, organized structurally: ranging from the direct consciousness of others in face to face encounters to the modes in which institutions are sedimented in deep historical time, and in which social interaction is carried on across broad areas of global space." Giddens (1983: 53)

Giddens (1991) returns to the theme of mediated experience when he talks about modernity and self identity.

> "In high modernity, the influence of distant happenings on proximate events, and on the intimacies of the self, becomes more and more commonplace. The media, printed and electronic, obviously play a central role in this respect. Mediated experience, since the first experience or writing, has long influenced both self-identity and the basic organization of social relations. With the development of mass communication, the interpenetration of self-development and social systems, up to and including global systems, becomes more and more pronounced. The 'world' in which we now live is in some profound respects thus quite distinct from that inhabited by human beings in previous periods of history. It is in many ways a single world, having a unitary framework of experience (for instance, in respect to basic axes of time and space), yet at the same time one which creates new forms of fragmentation and dispersal. A universe of social activity in which electronic media have a central and consecutive role ... " Giddens (1991: 4-5)

Giddens views modernity as "inseparable" from its own "media": the printed text, and subsequently, the electronic signal. Giddens states that the expansion of modern institutions was directly bound up with the tremendous increase in the mediation of experience which these communication forms brought in their wake.

Claude Fischer's (1992) work concerning the relationship of technology to society is similar in its belief in the dynamism between technology and society. Fischer, in asserting a social constructivist framework states "Mechanical properties do not predestine the development and employment of innovation. Instead struggles and negotiations among interested parties shape that history. Inventors, investors, competitors, organized customers, government, media - conflict over how an innovation will develop. The outcome is a particular definition and structure for the new technology." Contrary to Noble's (1977) deterministic paradigm, Fischer instructs us that new technologies may have second or third order effects that are unintentional, and that such unintentional consequences may contradict or limit other goals and purposes. The struggle for power and the centralization of control that telegraphy wrought are central themes in this work. Additionally, much attention is given to the way the telegraph either directly or indirectly mediated social and working relations within society.

This work is divided into two parts. Part I deals with the origin and diffusion of new technologies. Various academic disciplines have developed models to account for this process. Chapter 1 examines the various models in relation to telegraph technology and within a sociological framework. In this work, technology innovation is conceived of as a dynamic social process. It is the result of the interaction of scientists, citizens, social, cultural, and institutional forces. Merton's concepts of singletons and multiples and push/pull economic theory will be examined in the discussion of the origin of new technologies. The work of Rogers, Gold, Robertson, Markus, and Markusen will be discussed in relation to the invention and diffusion of new technologies. The final section presents a new model for the process of technology diffusion - one that is specific to interpersonal communication technology. Chapter two is an illustration of this new model using the telegraph as a case study and set within the innovation process as modeled by Rogers (1983), with a few modifications added. This chapter also looks at the inter-industry and inter-institutional linkages that aided in and were essential to the diffusion of telegraphic technology. The emphasis in this work on the relationship between technology and increased control, of which communication is an essential element, made it seem natural to examine the first radical innovation in interpersonal communication technology. The specific label of interpersonal communication technology is chosen to refer to those technologies which require more than one person at a time to adopt, and depend on the existence of a systemic infrastructure in order to be effective. This is in opposition to other communication technologies which do not require a systemic infrastructure of the industry in order to be diffused, for example, a radio or the newspaper. Interpersonal communication technology implies a priori knowledge about the party one wishes to communicate with. It involves person to person contact as mediated through a particular technology. This is in contrast to other forms of communication technology which involve person to information contact via a particular technology.

Part II of this work deals with impact of new technologies. A systems approach is taken in this section. In the terms of system analysis, an innovation will have a ripple effect beyond its intended point of impact. Telegraphy—a technology to improve communication — did in fact improve the speed at which people could communicate, as well as mediating the effects of distance. However, telegraphy also affected other institutions within society. In attempting to understand the impact of a technology one must go beyond the innovation itself. In the process of diffusing, a technology becomes part of that culture and alters relations within it (that is, the ways people interact with each other, the way they think, and the way they work). Technology often affects the organization of work and the workplace and can set a precedent for the future. Chapter three looks at the relationship between the telegraph and formal organizations. It examines the role telegraphy played in facilitating the growth of the modern business bureaucracy, as well as modern monopolies. The Western Union Telegraph Company will be discussed as the prototypical interpersonal communications firm.

The telegraph and the political system: power and politics form the subject matter for chapter four. The political debate concerning the ownership of this new technology stamped the development of telegraph technology as well as that of future communication technologies. And as a communication device, the telegraph was prototypical in its ability to be used to further political ends (propaganda, information control, censorship, national security). This chapter concludes with a discussion of two diffusion agencies which were important to telegraphy: the military and the newspaper.

Chapter five looks at the relationship between the telegraph and culture. The impact of the telegraph on cultural hegemony is explored, as well as some of the more novel social uses. Chapter six will focus on the telegraph and social stratification. Gabler 's (1993) ideas on the creation of a new middle class due to telegraphy are explored as well as the impact the telegraph may have had on the distribution of wealth and power.

The effects of telegraph technology also reach into the legal system. The history and function of patent law is important to the discussion of technological innovation. In Chapter 7, the evolution of patent law is used to illustrate the way in which control over innovation begins to shift from the inventor to the corporation with a rise of communication monopolies. It also demonstrates how the nature of patent law changed from one that was originally designed to protect the individual to a mechanism of control, and hence, power, for corporations. Additionally, the area of contract law and other novel uses of telegraphy within the judicial system are explored.

Chapter eight looks at the relationship between the telegraph and community. It shows how this invention facilitated an increased sense of nationalism. This chapter also explores the concept of the global community and the data superhighway. Records from Western Union indicate that the concept of the data superhighway, that is, of an integrated, multiple access network, was present as early as the 1960's. A discussion of Western Union's INFO-MAC Service and a diagram of Western Union's integrated record communications network are included in this chapter.

Acknowledgments

I am grateful to the many people who have interested themselves in this project. Especially helpful have been Dean Savage, Paul Attewell and Paul Montagna. I am thankful for having the privilege of knowing and working with the late Sidney H. Aronson who was an inspirational teacher, mentor and friend. I am particularly indebted to Frank Hull for his invaluable suggestions and guidance. To Margaret and Anthony Fiore, my first teachers, a special note of thanks. I am truly grateful for the constant support and encouragement of many friends and colleagues. Most essential too, were my husband, Gerard Lubrano, and our daughters Danielle and Valerie. I am most grateful for their constant love, patience and support throughout the many years this work was in progress.

A special note of thanks to Tania Bissell and Chuck Bartelt at Garland Publishing for their help and guidance.

The Telegraph

Part One

I
Origin and Diffusion of New Technology

Technology innovation is a process in which origin and diffusion are two of the major elements. Innovation implies change; embedded in the concept of social change is the question of the origin and diffusion of new technologies. Various academic disciplines have developed models to account for this process. This chapter will examine these models in relation to telegraph technology and within a sociological framework. Each of these models focuses on the diffusion process and tries to explain the S-shaped curve that usually results from plotting diffusion rates, primarily in terms of a single variable, for example, profit cycle or adopter characteristics. Each model works well for the particular kinds of technology that it deals with. However, the needs and nature of interpersonal communication technology do not make it amenable to any of the existing models. Consequently, an alternative model of technology diffusion is offered that is specific to interpersonal communication technology and which encompasses multiple variables as essential to driving the diffusion process.

ORIGIN OF NEW TECHNOLOGIES

Invention generally refers to the initial discovery of something new. Innovation is an activity. It introduces invention into the social structure, whether it be a new method for performing a task, a new custom, or a new product/device/service. Various theories have been put forth to account for the origin of inventions/innovations. Parker (1974: 36) posits three theories for invention:

Transcendentalist theory—gives primacy of place to individual genius. In this theory, the lone inventor makes a great impact with a single creative thought.

Mechanistic theory—invention is the offspring of necessity; need dictates and technology complies. In this theory economic forces predominate and the role of the individual genius is rejected.

Cumulative Synthesis theory—asserts that invention arises out of the accumulation of knowledge. Insight is required (this is where the individual plays a role) but one person's work is inseparable from the work of his/her predecessors.

Transcendental and cumulative synthesis theory both seek to stress human agency in the invention process. Both place the source of invention within the individual, though from dialectically opposed points of view. Mechanistic theory seeks to locate the source of invention within the economic institution of society. In this view, human beings are tools of the economic order. It is the needs of the economy, not merely the creativity and intellect of human beings (or collections of human beings), that drives the innovative force. It is not the human endeavor of gaining increasing control over life that dominates, but the survival and proliferation needs of the economic system that dominate. The theories Parker presents can be used as a framework within which to view the work of other researchers on the origin of new technologies.

Transcendentalist versus Cumulative Synthesis Theory

The opposition of these two theories represents the classic debate concerning the origin of new technology. Though the specific labels may change, the central question remains the same: Is innovation the result of individual genius, or is it the result of insight applied to a culture's accumulated store of knowledge? There is a strong temptation within American culture to recognize individual genius as this is in keeping with one of the basic values of American Culture—

individualism. Deeply rooted in our culture is the belief that success or failure is primarily within the control of the individual due to his/her characteristics, abilities, and motivation. Consequently, formal recognition for achievement is necessary as a hallmark of success. We are not a society oriented to collectivism and thus theories of the cumulative synthesis nature are not emphasized in our history books but do bear examining if we are to acquire a full understanding of the innovation process.

Singletons and Multiples

Merton's (1974) work falls under the cumulative synthesis theories. Merton addresses the question of the origin of innovations and frames it in terms of the battle between the social determination theory of discovery (that is, that the course of science and technology is a continuing process of cumulative growth) and the heroic theory (that is, the result of men of genius, who, with their ancillaries, bring about basic advances). Merton posits that we are often misled into believing that individuals are responsible for scientific innovation because of the practice of eponymy. He offers as examples Boyle's Law and Planck's Constant, and one could easily think of other examples. Merton (1974: 215) writes that this practice leads to "the virtual anonymity of the lesser breed of scientists whose work may be indispensable for the accumulation of scientific knowledge—these and similar circumstances may all reinforce an emphasis on the great men of science and a neglect of the social and cultural contexts which have significantly aided or hindered their achievements." In the dynamic view of innovation origin and diffusion, innovation comes about due to the work of various scientists, their ability to draw from the accumulative culture, as well as from social need. Citing the work of William F. Ogburn and Dorothy S. Thomas concerning the multiple and independent appearance of the same scientific discovery, Merton (1974) concludes that such innovations became virtually inevitable as certain types of knowledge accumulated in the cultural heritage, and, as social needs diverted attention to particular problems. Consequently, one can conclude that the confluence of these events will lead to the multiple and independent appearance of a scientific innovation. Merton (1974: 356) states that "the pattern of independent and multiple discoveries in science is in principle the dominant pattern rather than a subsidiary one. It is the singletons—discoveries made only once in the history of science—that are the residual cases, requiring special explanation."

Both Parker and Merton agree that the lone inventor theories are not adequate to explain invention and innovation. They see theories that account for the dynamics of need and accumulated culture as providing a more accurate picture. Telegraphic history can serve as an example in support of the cumulative synthesis type theories, and Merton's theory of multiple invention in particular.

The discovery of electricity was the innovation which would ultimately lead to the first modern innovation in communication technology. This knowledge became part of the accumulative cultural base of many Western European countries (as well as America after its settlement) and men of science all around the world, were searching for a way to apply electricity to communication. They were addressing the problem of how to send electrical impulses through wire. George Louis Le Sage, Francis Ronalds and others experimented with lines to conduct electricity. They needed the work of Galvani, Volta, Oersted and Ampere before science had the practical knowledge to apply electricity to long distance. It was the combined work of Arago, Sturgeon, Steinheil, Weber, and other European scientists that Samuel Morse built upon to "invent" his electro-magnetic telegraph. (Much scientific knowledge was shared through the papers scientists published, as well as through informal, salon-type discussions). Though the groundwork for the invention of the telegraph was well laid, no telegraph proper (that is, no instrument that could "write" messages at a distance) had been invented until Morse did so. Another similar device was developed around the same time (1830's) by the Englishmen Cooke and Wheatstone. Their telegraph operated on a different principle from Morse's—it did not leave a written record, but it does help to explain how England had an operating telegraph only shortly after the United States. Table 1-1 depicts a history of the electric telegraph and clearly illustrates two important facts: that though Morse's name is most closely associated with the discovery of the telegraph in America, the same discovery was independently and simultaneously being made in England; the invention of the telegraph was the product of an accumulated store of knowledge. The previous discoveries of other scientists were of crucial importance. The dissemination of knowledge and its incorporation into the cultural base are important components for the multiple and independent discovery hypothesis. It should be noted however, that the very dissemination and incorporation of such knowledge brought about one of the earliest fights over " intellectual property" rights. A discussion of the impact of this argument can be found in chapter 7 in the section on patent law. The words of Dronysius Lardner (1855: preface) seem apropos here: "The Electric Telegraph is not the invention of an individual . . . it is the joint production of many eminent scientific men and distinguished artists of various countries, whose labours and experimental researches on the subject have spread over the last twenty years."

Telephone technology has a similar history of simultaneous development, only in this case, the simultaneous discovery was by two Americans. Both Alexander Graham Bell and Elisha Gray "invented" the telephone. Both filed for patents on their telephone devices on February 14, 1876. Building upon the work of other scientists, and driven by the desire to transmit sound over the telegraph wires (but for different reasons—Bell was seeking to aid the deaf, Gray to transmit music), both men discovered a means by which human voices could be transmitted over the wires. In addition to the work of the scientists who contributed to the invention of the telegraph, the work of Herman Von Hemholtz, Thomas Edison, and the researchers at The Massachusetts Institute of Technology was crucial to the invention of the telephone.[2] Again we see that invention was the product of need and insight applied to accumulated knowledge.

Telegraph and telephone technology are two illustrations of Merton's theory of multiple invention as the result of social need and accumulated cultural knowledge. However, each highlights a different aspect of the idea of social need. Telegraph innovation grew out of the general need for greater control over information processing, and was made possible by Morse as well as Cooke and Wheatstone as they built upon the knowledge available in their respective cultural heritage's. Telephone innovation is an example of innovation growing out of a more particular sense of need, for example, Bell's desire to aid the deaf. Though Bell's invention didn't satisfy the particular need he was driven by, it does illustrate how social attention can be focused on an area of study (that is, the instantaneous transmission of information) which leads to varied inventions.

Table 1-1. History of the Electric Telegraph

	History of the Electric Telegraph[1]
1600	William Gilbert published De magneta, in which he coined the word "electric" from the Greek elektron (meaning amber), and laid the foundations of the science of electricity by collecting together various scattered observations on how a piece of amber, when rubbed, acquired the property of attracting small pieces of straw or paper, and adding his own observations
1660-1663	Otto von Guericke invented the first machine to generate electricity.
1675-1751	Various frictional machines were developed, so that by the middle of the 18th century, the frictional machine was established as a fairly reliable generator of electricity.
1729	Stephan Gray laid the foundations upon which the electric telegraph could be built with his discovery of the basic principles of conduction and insulation. Gray "succeeded in transmitting electric charges to a distance of nearly 300 yards, and it is of interest to note that, in so doing, he had in fact set up all the essential elements of the electric telegraph: a source of electricity at the sending-end, an insulated line, some form of indicator at the receiving-end (Gray used a down feather), and an earth-return circuit . . . " (p.649)
	Telegraphs Employing Static Electricity
1753	C.M.'s (identity unknown) letter appears in *Scots Magazine*, giving detailed specifications for an electric telegraph using static electricity.
1782	Le Sage develops a prototype of C.M's telegraph.
1787	M. Lomond gave a demonstration in Paris of a practical experiment of an electric telegraph using static electricity which was a refinement of C.M.'s specifications.
1795	Don Francisco Salva of Barcelona demonstrated a multi wire scheme for telegraph, as well as a scheme for submarine telegraphy.

Table 1-1. History of the Electric Telegraph (continued)

1816	Francis Ronalds demonstrated a telegraph in which he connected a friction machine, which charged the wire, to an indicator. A dial was marked with the letters of the alphabet and rotated behind a window-like opening, located at both the sending and receiving ends, and the dial were rotated synchronously. The charge was stopped when the desired letter appeared in the window.
	Electrochemical Telegraphs
1791-1800	Galvani and Volta advanced the knowledge of the field of electricity especially the relationship between electricity and chemical action. This resulted in "a Variety of electrochemical telegraphs." (p.653)
1809	S.T. von Soemmering invented a telegraph similar in principle to Salva's, but with the addition of an alarm to call the attention of the receiving operator.
	Electromagnetic Telegraphs
1820	Hans Christian Oersted published his observation that a magnetic needle was deflected when a current of electricity flowed through a wire above it.
1820	Ampere and Aragon in France; Humphrey Davy, Faraday and Sturgeon in England; Schweiger and Schilling in Germany; Henry in America all enhanced Oersted's work on the magnetizing effect of an electric current, and experimented with applying this knowledge to telegraphy.
1837	Cooke and Wheatstone in England, granted their first patent for a five-needle telegraph
1838	With the help of Joseph Henry and J.D. Gale, Samuel F.B. Morse introduced his telegraph using the dot-dash system of communication, which later became the industry standard.
1844	The formal use of the Morse telegraph began.
1866	Beginning of international submarine telegraphy with the successful laying of the transatlantic cable.

Mechanistic Theory

Mechanistic theory gives primacy of place to economic factors. The language and framework of mechanistic theory is that of the economists. Technology is the response to market demand whether it be industry driven or consumer driven.

Push/Pull Economic Theory

A specific application of the mechanistic theory of invention is push/pull economic theory. According to this theory, the sources of innovational pressure are dichotomized as: push—innovation as the result of the initiatives of research and pull—innovation as the result of the expression of customer/consumer need. Using the history of early telegraphic technology in Europe as an example we can illustrate the "pull" aspect of this theory. The immediate predecessor of the electric telegraph was the optical telegraph, or semaphore. This telegraph was invented by Claude Chappe, a Frenchman, in 1794. Chappe wanted to devise a system of communication that would enable the central government to receive and transmit intelligence information and orders in the quickest possible time. The French government, at war with most of its neighbors, adopted this form of communication during the French Revolution. The expansion of the telegraph system in France continued even after Napoleon seized power in 1799. By 1844, France had 533 stations and 5,000 kilometers of line. The semaphore system that Chappe invented eventually spread all over Europe. The British developed their own version of the telegraph after noticing the way the French were communicating, and after finding a drawing of Chappe's telegraph they set out to produce a better system. Garratt (1967: 646) posits "In England, reports of the working of the Chappe line between Lille and Paris were received during the autumn of 1794, and an illustrated description appeared in *The Sun* on the 15 of November. Although it was not uncommon for similar inventions to be made quite independently yet simultaneously in widely different places, it seems likely that it was those reports from France that stimulated Lord George Murray and John Gamble to propose schemes for visual telegraph systems to the Admiralty in 1795." In the late 1790's Britain built telegraph lines all along its coast. Among the other European countries, Sweden was quickest to set up a telegraphy system of their own in 1795 due largely to the work of Abraham Edelcrantz. The Germans had a primitive form of the semaphore in 1798, and Denmark had a modified form by 1802. [3] However, when military need ebbed, the optical telegraphs tended to be abandoned because they required extensive manpower and expense to operate. Stations were seldom more than eight miles apart and were subject to interruption due to adverse weather conditions. These systems were of little use to the individual citizen or to the business community in Europe at the time.

This example of "pull" theory illustrates an important point in relation to the innovation process. Technology that is developed in response to a particular need has a guaranteed pool of potential adopters and is therefore more likely to display a quicker beginning in the diffusion phase of the innovation process as seen in relation to the shape of the S-curve.

Transcendental Theory Versus Cumulative Synthesis and Mechanistic Theory

Cumulative synthesis theory and mechanistic theory are not incompatible. They can work in concert with each other when one examines the origin of innovation from an organizational perspective. What emerges from this perspective is a conflict between the idea of the lone individual inventor versus innovation as the offspring of formal research and development departments within formal organizations.

Individual Versus Research and Development Within Organizations

The controversy between heroic and cumulative synthesis theories can be cast in yet another form—individual research versus research and development within and organization (transcendental theory vs. cumulative synthesis and mechanistic theory). According to this debate, the individual inventor, motivated by dreams of personal acclaim and financial success, is thought to be the driving force behind innovation. Conversely, the opposition view states posits that it is the accumulated knowledge within the formal research and development departments of organizations that is the driving force behind innovation, motivated by economic need. Research can be found to support both positions. For example, in support of the independent inventor is the work of Charpie (1970) who informs us that among industrialized economies 30%-50% of long term economic growth stems from innovations that improve productivity. Another 30%-50% of long term growth stems from innovations which lead to new products, processes, or completely new industries. With such a high percentage of long term growth dependent on innovation, it only seems logical that companies should invest part of their money in research and development departments. In fact, large companies in the United States spend considerable amounts in research and development. Charpie (1970: 6) further states: "In the United States of America over 80 per cent of the Research and Development dollars are spent in two hundred large companies." Yet, he goes on to inform us, that during the twentieth century a "remarkable percentage" of the important innovations were made by isolated individuals working within very small firms. Charpie cites the studies made by Jewkes in England, Hamburg in Maryland, Peck at Harvard, and Enos at the Massachusetts Institute of Technology, all of which conclude that more than two thirds of the basic discoveries which led to important innovations during this time period were from independent inventors or small firms. The FM radio, the dry electrostatic copier, penicillin, streptomycin, the zipper, the modern rocket engine, instant photography, air conditioning are all examples of the types of innovations which were *not* products of research and development (R & D) departments within large firms. Charpie explains this phenomenon by referring to the values and traditions of American culture, that is, American individualism coupled with the American tendency to judge success by financial standards. Thus, Charpie (1970: 18) instructs us that entrepreneurship leads to entrepreneurship because people are motivated by the success and high visibility of technical entrepreneurs who have "achieved personal and financial success". Goldman (1970: 12) also espouses the individual entrepreneur as the main source of innovation. He believes that of all those innovations that have led to economic expansion, " . . . it is the single individual champion, the technical entrepreneur, who carries the ball as it were and pushes the technology ahead". One can clearly see the association here with the heroic theory of invention.

The opposing position can be represented in the work of Pavitt (1970) who believes in the necessity of strong research and development departments within large firms, as the important source for innovation. Pavitt informs us that the existence of R & D departments within large firms creates the possibility for scientists and engineers to start up their own firms. He says that large firms supply the knowledge base for scientists and engineers, who then leave the firm in order to apply and exploit commercially the knowledge they have acquired. In other words, the R & D departments of large firms serve as the training ground for, and the initial capitalization of the innovations of individual independent innovators. Clearly there is a belief in the need for organized, ongoing basic research. Mervis (1994) informs us that even in these tight financial times the government is not cutting R & D spending. It plans to continue funding with the hope that these new technological innovations will help keep the economy going.

This debate also serves to point out one of the conceptual problems in the issue of the origin of innovation, which is that innovations cannot be treated as if they were all alike in character. A distinction needs to be made between radical innovation and incremental innovation. Radical innovation refers to the introduction of something entirely new, that is, not previously available or possible. Incremental innovation involves the modification or redesign of a product or process so as to result in higher quality, additional performance capabilities, enhanced performance, lower cost, or any combination of these. Research has shown that innovations that lead to the improvement of a product/process usually come from within the research and development departments of large firms. New product/process innovations or directions often stem from the independent entrepreneur or small firm. Mueller (1964) informs us that large companies that enjoy a pre-eminence in their field are more successful in making product/process improvements than in discovering new products. The history of communication technology can be used to exemplify this idea.

The telegraph was initially the product of individual inventive effort. However, once the telegraph entered into its diffusion phase, it was eventually subsumed within a formal organization, The Western Union Telegraph Company, which quickly became the dominant entity in the telegraph industry. Hence, after the initial period of radical invention, telegraph innovation evolved into a case of organized incremental invention because concern shifted to the maintenance of the company's dominant position within the telegraph industry. In order to maintain its control over the rapidly growing industry Western

Union established the Department of the Electrician in 1873. It was from among the ranks of these engineers that further improvements and innovations to telegraphy came. As this department evolved within the firm, credit for particular innovations was shifted from the individual "inventor" to the department as a whole. For example, an internal Western Union memo entitled "Outline of the History of Quotation Tickers", depicted in Table 1-2, illustrates this shift.

Table 1-2. Outline of the History of the Quotation Ticker

Ticker	Inventor	Date
Calahan 3-wire Escapement	Edward A. Calahan	1868
Universal 2-wire Escapement	Thomas A. Edison	1870
Phelps 2-wire Escapement	George M. Phelps	1870
Manhattan 1-wire Weight	Charles T. Chester	1871
Phelps 1-wire Escapement	Henry Van Hoevenbergh (adapted from Phelps 2-wire)	1874
Phelps 1-wire Weight	George M. Phelps	1876
N.Y. Quotation 2-wire Weight	Stephen D. Field	1880
Scott 2-wire Weight	George B. Scott	1883
Scott 1-wire Weight	George B. Scott	1885
Burry 1-wire Self-Winding	John Burry	1890
Scott-Phelps-Barclay-Page Self-Winding	Adapted from Scott 2-wire	1903
Western Union Automatic Self-Winding	W.U. Engineers	1923

D'Humy and Howe (1944: 1) state that telegraphy " . . . was the real beginning of electrical engineering" in as much as it was the real beginning of the application of electricity to the benefit of mankind.[4] Consequently, we see a shift in power and control in society, from the individual to the organization, increasing the dominating position of business in the then emerging, capitalist industrial America.

Noble (1977) tells us that invention became institutionalized; increasingly patents became the property of corporations rather than of the individual inventor. He further tells us that the solitary inventor had two choices in the face of the development of large scale corporations: fight for their rights and seek to develop their invention on their own, or, form an allegiance with a corporation—selling their patent rights in exchange for employment security, thus also ensuring the financial backing for further research and development. This shift in the relationship between inventor, patent holder, and corporation meant that emerging corporate America gave birth to the idea of the corporation as inventor (as Noble 1977, labels it). The power that this gave to corporations at the expense of the individual is not be denied. Though Lind was writing about the 20th century in his article "You Can't Skin a Live Tiger", his writings seem equally applicable to the era of telegraph and telephone technology. Lind (1949: 109) writes "The problem we face today is that, in an era that increasingly lives by science and technology, business control over science and its application to human needs, gives private business effective control over all the institutions of democracy, including the state itself." For, what was just emerging was the "wedding of science to the useful arts". Noble (1977: 5, 7) states that "Modern science-based industry—that is, industrial enterprise in which ongoing scientific investigation and the systematic application of scientific knowledge to the process of commodity production have become routine parts of the operation—emerged very late in the nineteenth century" and that " . . . it was they [the men of the period between 1880-1920 who created and ran the modern electrical industry] who introduced the now familiar features characteristic of modern science-based industry; systematic patent procedures, organized industrial-research laboratories, and extensive technical training programs." Certainly, one would have to credit telegraphy as being integral to this epochal wedding. Electrical and magnetic phenomena were originally the province of natural science, physics in particular. In Morse's quest to develop telegraphy he received aid from the physics laboratory of New York University in improving his original instrument. Subsequent to the successful "invention" of telegraphy, telegraph electricians began to meet informally to exchange views. These meeting eventually metamorphisized into a formal institution. An excerpt from *The Proceedings of the I.R.E.* (1944: 446) informs us that

> "In the United States, 21 telegraph men and four others joined to call into existence the American Institute of Electrical Engineers, the first two

> presidents of which were telegraph men. Many of the engineering departments of today's universities were founded by men who themselves studied electricity in college physics because of their early interest in telegraphy . . . [W]e electrical engineers, radio engineers, industrial electronics engineers, trace our ancestry back to Samuel Findley Breese Morse."[5]

Thus, in assessing the impact of telegraphy, we see that telegraphy was instrumental where patent ownership, organized industrial research, and the use of technical training programs are concerned. However, it is important to note that organized incremental innovation within science-based industry has some weaknesses, one of which is an inability to identify and respond to new radical innovations. The Western Union Telegraph Company can also be used to illustrate this narrowness of vision which can occur within formal research and development departments.

Western Union concentrated almost entirely on improving telegraphy, to the point of ignoring alternative developments. Western Union ignored the telephone until Bell started up his company. At that point, Western Union bought the patent rights of another independent inventor, Elisha Gray, who was also working on developing the telephone. Western Union was in the telephone business from 1876 to 1879. At the same time the National Bell Telephone Company was vigorously developing its business. The Bell Co. claimed exclusive rights to the telephone field because of its patents (see the section on patent law in chapter 7 for a fuller discussion of the relationship between the communication field and patent law) and many suits and countersuits were filed between Western Union and the Bell Company. A document from the Western Union archives entitled "Supplementary Report on the Outline of A.T.&T. Co. Relationship to Telegraph Business" written in 1933 states "Both parties sought a mutually satisfactory settlement. Few men then had any conception of the ultimate scope of telephone business. Western Union was concerned largely with keeping the Bell Company out of the telegraph field." In November of 1879, Western Union and the National Bell Telephone Company made a seventeen year agreement representing a compromise. This same document informs us that "The principal features are carefully framed provisions for the protection of the Western Union telegraph system, and a lease and transfer by it to the Telephone Company, of Western Union interests in telephonic patents, its telephones and telephone exchanges. This agreement also provided for Western Union to receive a certain proportion of rentals or royalties which should come to the Telephone Company. Western Union viewed this contract with considerable satisfaction."

Maclaurin (1964: 73), writing on the process of technical innovation, states that Western Union's management was more interested in buying up competitors and making protective agreements than in the fundamental development of communication, and this ultimately hurt Western Union. "The telegraph industry, however, had failed to visualize the potential importance of the telephone, and by 1900, was beginning to feel the competition of this alternative method of communication which it could have controlled." Maclaurin further states (in support of Mueller's idea) that among monopolistic enterprises there is no genuine interest in radically new products.

This same pattern is also evidenced in the history of the Bell Telephone system. Innovations primarily took the form of improvements within the industry at the expense of new products. For example, the Bell system offered only a basic black phone for much of its history. Yet, in the face of competition, it finally got into the designer phone market. And, we can see a similar state of affairs in the relationship of IBM to the personal computer. This innovation came from outside IBM and IBM was definitely a late comer to this market.

The preceding section also helps to identify another factor which is useful in charting the trajectory of a new invention, that is, the distinction between radical and incremental inventions. The history of interpersonal communication technology as exemplified by the telegraph suggests that radical inventions are more likely to come from the "individual" inventor and hence are less likely to have a guaranteed pool of potential adopters, resulting in a slower beginning to the s-curve for that invention. Conversely, incremental inventions are more likely to come from within formal research and development departments within formal organizations. These inventions are more likely to have a guaranteed pool of potential adopters in that the inventions enhance technologies that have already been diffused.

Consequently, one could expect to see a steeper beginning to the s-curve for such an incremental invention.

DIFFUSION OF NEW TECHNOLOGIES

The prior sections discussed theories of the origin of technologies, but equally important is the acceptance and diffusion of a new technology. It is generally recognized that there needs to be a wedding between the enunciation of a new idea in science and its practical application in order for innovation to take place. In other words, a link must be forged between scientific knowledge and commercial success. This idea was expressed by Professor Leonard Gale on January 22, 1872 when he stated in a letter that "Men of science regard the discovery of a new fact in science as a higher attainment than the application of it to useful purposes, while the world at large regards the application of the principle or fact in science to the useful arts as of paramount importance. All honor to the discoverer of a new fact in science; equal honor to him who utilizes the fact for the benefit of mankind." The diffusion process serves as this link. Diffusion is the process by which the use of an innovation spreads and grows. Adoption and imitation are important components which help spread usage from the source.

The process of diffusion of innovation has been modeled by various researchers. The work of Rogers (1983), Robertson (1981), Markusen (1987) and Markus (1987) will inform the present study. Different paradigms of innovation and diffusion highlight different dominant processes. Rogers (1983: 10) defines diffusion as " . . . the process by which (1) an innovation [6] (2) is communicated through certain channels (3) over time (4) among members of a social structure. Rogers' model can be considered representative of the general orientation in social science (sociology and anthropology) to the adoption model of innovation diffusion. This perspective emphasizes the informational, communicative, and cultural aspect of diffusion. It suggests that diffusion rate is dependent upon an innovation's congruence with the personal characteristics and norms of a population. Adoption models are concerned with the decision-making process, social networks, informational flows, and focuses on the process by which adoption occurs. Consequently, one could say that the adoption model represents the "pull" aspect of innovation and diffusion, and this research has focused primarily on the diffusion of consumer oriented innovations.

Rogers breaks down the innovation development process into six phases, which, he tells us, do not necessarily have to occur in the order listed, and some may be skipped altogether for certain innovations. Rogers' six stages are depicted in Table 1- 3. Table 1-4 depicts the stages in the innovation decision-making process and Table 1-5 is a typology of innovators based on time of adoption. This is the diffusion model that Rogers uses to explain the s-curve for technology diffusion.[7] Examination of these tables makes very clear the importance of the individual and communication in the adoption models.

Table 1-3. Rogers' Six Phase Innovation Model

Phase One—Needs and Problems	Recognition of a problem or need stimulates research to solve the problem or need.
Phase Two—Basic and Applied Research	Basic research—investigations for the advancement of scientific knowledge. Applied research—investigations that are intended to solve practical problems. Invention is the result of the dynamics between basic and applied research which leads to development.
Phase Three—Development	Putting a new idea in a form that is expected to meet the needs of an audience of potential adopters.
Phase Four—Commercialization	Production/manufacturing/packaging/ distribution of the embodiment of an innovation
Phase Five—Diffusion and Adoption	Involves decisions concerning technology gatekeeping, that is, which innovations should be diffused, trials to determine the effects and efficiency, and finally the decision to diffuse.
Phase Six—Consequences	Original need/problem is either solved or not solved by the innovation.

Table 1-4. Stages in the Innovation Decision Process

Stage 1 Knowledge	Occurs when an individual or other decision-making unit is exposed to an innovation's existence and gains some understanding of how it functions.
Stage 2 Persuasion	Occurs when an individual or other decision-making unit forms a favorable or unfavorable attitude toward the innovation.
Stage 3 Decision	Occurs when an individual or other decision-making unit engages in activities that lead to a choice to adopt or reject the innovation.
Stage 4 Implementation	Occurs when an individual or other decision-making unit puts an innovation into use.
Stage 5 Confirmation	Occurs when an individual or other decision-making unit seeks reinforcement of an innovation-decision already made, but he/she may reverse this previous decision if exposed to conflicting messages about the innovation.

Table 1-5. Diffusion Model—Typology of Innovators

Innovator	Venturesome; Plays the role of launching a new idea into the social system
Early Adopters	Respectable: plays the role of opinion leader; role model for many other members of a social system; deliberate
Early Majority	Deliberate; important link in the social systems networks
Late Majority	Skeptical; adoption may be an economic necessity and/or an answer to increasing network pressure
Laggards	Traditional; adoption lags far behind awareness or knowledge and slows down the innovation-decision process

Economists, when doing diffusion research, have tended to focus on the speed of innovation spread, the impact of innovations on industries in terms of profitability and growth, and the impact of management structure and decisions. The market and infrastructure paradigm informs much of the research of economists on diffusion innovation and tends to focus primarily on technological innovation on the firm or industry level. Spatial aspects of diffusion, diffusion agency, and infrastructure are important and dominant factors in this perspective. The focus on diffusion agency[8] rather than on the adopter as critical in the diffusion process represents a shift to considerations of the "supply" aspect of diffusion as primary. The models of diffusion presented by Robertson (1981) and Markusen (1987) can be used to illustrate the economic perspective.

Robertson (1981) posits a product life cycle model and discusses innovation diffusion as a series of stages. These stages, depicted in Table 1-6, are used to explain the S-shaped curve which occurs in plotting diffusion as the rate of change of a performance parameter over time. Robertson asserts that marketing is crucial to diffusion. Robertson also instructs us that one must consider the reactions to innovation when exploring diffusion and notes that factors that impact on businesses' reactions (such as patent law, labor, profitability) are different from those that impact on individuals (want, social status, satisfaction, one upmanship). Both must be taken into account when examining the diffusion of an innovation. Robertson (1981: 105) writes: "The skill in marketing lies in identifying potential customers and employing the most effective means to awaken perception and effective demand."

Table 1-6. Robertson's Stages of Innovation Diffusion

Introduction	Period of slow market response to a new product or process.
Growth	Period during which early reluctance gives way to quickening adoption.
Maturity	Period during which the cumulative proportion of users reaches its peak.
Decline	Period during which sufficient users are switching to alternatives or giving up use altogether; curve falls below the level reached in the previous stage.

Markusen (1987), believing that the product life cycle highlights the pattern of output while neglecting the motivation or behavior of the decision maker, offers yet another model—the profit cycle model (depicted in Table 1-7). This model hypothesizes the constitution of sequential temporal stages in the diffusion process which correspond to the development of an industry and to different behaviors at each stage oriented to profit.

Table 1-7. Markusen's Profit Cycle Model

Zero Profit - Experimentation Stage	This is the stage of research and design experimentation. Inventive activity is aimed at inarticulated social/corporate needs or bottlenecks that hamper the profitability of contemporary production.
Superprofit—Innovation Stage	After successful innovation, there is a dramatic growth stage in which profits rise well above the economy-wide level. These superprofits are a return that the entrepreneur/corporation is able to command in a market that it dominates in the short-run. During this period average cost goes down due to standardization.
Normal Profit—Competition Stage	The entry of new firms into the market chips away at superprofits. Growth slows as supersaturation is reached. Smaller firms usually go out of business during this stage.
Normal-Plus and Normal-Minus Profits—Maturity	Normal profit continue unless new products enter the market to challenge that sector. Innovative drive ensures the obsolescence of most products/processes. Two responses are possible to this situation: oligopolizing (domination of the market by a few sellers, which leads to price increases); decline in the face of competition from substitutes or imports.
Negative Profits—Decline	This is the stage of absolute losses. Plants will close, capacity will be sold to another user.

The "critical mass" theory of interactive media suggests a model for the diffusion of interactive communication technology specifically. M. Lynne Markus (1987) suggests that interactive media have two characteristics that are not shared by many other innovations: widespread usage creates universal access;[9]reciprocal interdependence.

Universal interdependence is a public good that individuals cannot be prevented from enjoying even if they have not contributed to it.

(Markus 1987:491) Critical mass theory[10]seeks to predict the probability, extent, and effectiveness of group action in pursuit of public good. Two variables are important in this theory: shape of the production function (accelerating or decelerating); heterogeneity of resources and interest in the population (Markus 1987: 496) Markus informs us that the general shape for the production function for interactive media is accelerating (that is, successive contributions generate progressively larger payoffs, thus making additional contributions more likely). Furthermore, Markus tells us that "the primary resource individuals contribute to the collective outcome of universal access is their readiness to reciprocate communication (p. 499). Other resources, such as operational access and information that can be exchanged, and/or interests, such as an active communication consumer or partner accessibility, are also considered as contributions to universal access. Once a certain proportion of users (critical mass) have begun to use a new interactive media, use should spread rapidly throughout the community.

Markus' work seeks to explain develop a theory which will "explain the success or failure of universal access in a community into which an interactive medium has been introduced. The outcome of this theory (universal access) is a property of the community, but the antecedents are properties of individuals. Thus, the theory relates micro (individual) inputs to macro (community) outcomes, in contrast to research [Rogers, 1983] that explores the determinants of individuals' media choice behavior." (p. 493)

Though Markus' work focuses on the effects of group action as opposed to individual characteristics, as in Rogers'(1983) model of the diffusion of new communication technologies, both fail to take into account the systemic needs of the new technology from the perspective of building the physical infrastructure of that technology. One needs to look at the costs/benefits to the companies which own/control the new technology as well as the costs/benefits to the individuals or the community. Markus' work certainly suggests itself as an important factor, but not a lone factor, in attempting to explain the diffusion of communication technology.

The models presented here either implicitly or explicitly accept the major stages of the innovation process—needs, research, development, diffusion, consequences. The major differences lie in their elaboration of the diffusion phase of the innovation process. However, neither the adoption model which focuses on individual behavior at the point of adoption and diffusion and stresses the cultural infrastructure over the economic, nor the economic model which focuses on technology innovation as primarily an economic endeavor motivated by and oriented to profit and constrained minimally by the cultural infrastructure, are particularly adaptable as an explanation of the diffusion of interpersonal communication technology. Neither orientation can be given primacy of place in that both orientations play a major role in the diffusion of interpersonal communication technology but at different stages of the diffusion process due to the unique nature and needs of interpersonal communication technology. The collective nature and systemic needs of interpersonal communication technologies make them essentially different from other technologies which can be independently adopted and don't require a system in order to effect efficient diffusion. The fact that a particular communication technology is interpersonal implies that collective action and coordination are essential in the decision to adopt during the initial diffusion period. Consequently diffusion agency becomes an important variable in the early stages of diffusion for an interpersonal communication technology as these agencies are representative of collectivities (for example, the government or industry), giving access to and providing coordination among groups of people at a time. Additionally, one must consider the effects of historical or economic events which may motivate various diffusion agencies to adopt a new invention (for example, war, economic depression/boom).

Equally important is the distinction between radical and incremental inventions. If the invention is radical in nature, then the cultural infrastructure assumes great importance in that one will have to use the values of a given society as a framework in which to bring about normative change. For example, the case of the telegraph required a reorientation in the basic way people communicated with each other at a distance. Collectivities (groups of people in the case of the local district telegraph companies, which were subsidiaries of Western Union, as well as organizations) had to be convinced that this new form of communication afforded them more control over their lives than that which hitherto existed, or in other ways offered them the ability to meet a need that could not be met equally as well using some other means of interpersonal communication. Hence, recognition must be given to the dynamic interplay of cultural, economic and historical factors which can speed/retard the adoption of a radical new invention.

The systemic needs of interpersonal communication technology dictate the organizational structure for the industry that is to become the vehicle for the diffusion of that technology. Unitary control is essential in the building of a systemic network in order to obviate the problems that could arise if the physical structure were multiply owned. For example, what would happen to the continuity of the network if a company went bankrupt or failed to maintain its wires properly? The uninterrupted flow of information across the physical system is crucial for the proliferation of such a technology. In addition to building the physical system necessary for the transmission of a particular technology, one technical standard must become dominant, and hence we see the evolution of a natural monopoly is virtually inevitable for a radically new invention in interpersonal communication technology. This characteristic points to the reason why such technologies do not fit into the profit cycle model. Markusen (1987) posits a superprofit stage that comes from an industry dominating the market in the short-run. The competitive stage slows the growth of diffusion and chips away at the superprofits. Just the opposite seems to occur in the diffusion of interpersonal communication technologies. The emergence of competing companies during the initial phases of the diffusion process actually means slow growth due to the lack of standardization or of a system. Hence, the period of rapid economic growth for the industry that evolves as the vehicle for the diffusion of interpersonal communication technology is the period *after* the competitive stage.

It is also necessary to point out that for other forms of communication technology (both interpersonal and non-interpersonal) establishing a technical standard is also crucial. However, if these technologies depend on the infrastructure of a pre-existing interpersonal technology, than monopolization is not a necessity for successful diffusion. For example, fax and personal computer technology and the Internet rely on the pre-existing infrastructure of the telephone. The systemic needs of these technologies are thus already served by telephone technology. Consequently, successful diffusion rested more heavily on standardization.

Thus, I am proposing another model of diffusion that is specifically and purposefully designed to account for the diffusion process for one particular type of technology, that of interpersonal communication technology. Gold (1980) suggests that not all innovation should be treated as if it were the same thing. Communication innovation is different from manufacturing innovation which are both different from medical innovation, or transportation innovation, etc. Each has its own history and trajectory of development. Additionally, Gold (1980) and Brown (1981) suggest that a more over-arching or standardized framework may be desirable that can be applicable to any type of innovation. However, this can only be possible once we understand the differences in kind among the various technologies. By modeling the diffusion process for the telegraph and offering it as prototypical for all interpersonal communication technologies, I hope to identify elements that need to be taken into account when someone undertakes the task of extrapolating the general, multidisciplinary model from the various specific ones. The next section provides an explanation of the proposed diffusion process for interpersonal communication technology.

A DIFFUSION MODEL FOR INTERPERSONAL COMMUNICATION TECHNOLOGY

It is important to remember that diffusion is only one part of the technology innovation process. The diffusion model hypothesized here is placed within the general framework of the technology innovation model offered by Rogers (1983). However, certain elements have been modified and elaborated upon as they are of particular importance in understanding the diffusion phase as it relates to interpersonal communication technology innovation. Therefore the first part of this section reviews the technology innovation model in light of these changes (with the exception of the diffusion phase).The second part of this section is devoted to an explanation of the new diffusion model being offered.

Technology Innovation

Needs

Technology innovation is an intimate part of human history and can be viewed as attempts at manipulating the environment in order to gain greater control over it, as well as over life, and other people. Consequently, there is reason to break the concept of need into two elements—general, those arising out of a general concern for increasing control or human agency, and specific, those arising from the identification and enunciation of a particular problem. Each of these elements has a different potential impact on the innovation process and is hypothesized to affect the shape of the S-curve for the rate of diffusion of a particular technology. Technology generated by general need has an amorphous pool of potential adopters and hence would most likely have a flatter beginning to its S-curve. Technology developed to meet a specific need would have a base of guaranteed adopters suggesting a steeper rise in the shape of the S-curve.

Research

Basic research is often linked to general needs and applied research to specific needs. However, there is an interplay between these two types of research. Both are necessary, at some point in the history of a technology, to bring the enunciation of a new idea to practical application, that is to the invention stage. The sources of innovational pressure are diverse. It can come from need, or it can come from formal research and development organizations, or it can come from an individual.

Invention and Development

Not all inventions are alike and a differentiation needs to be made between the kinds of invention when assessing technology innovation. Kash (1989) suggests the labels of radical and incremental to differentiate between the different types of invention. Mokyr (1990: 295) proposes a similar distinction when he writes :

> "Microinventions generally result from an intentional search for improvements, and are understandable—if not predictable—by economic forces. They are guided by the laws of supply and demand and by the intensity of search and the resources committed to them. . . . Macroinventions . . . seem to be governed by individual genius and luck as much as by economic forces. Often they are based on some fortunate event, in which the inventor stumbles on one thing while looking for another, arrives at the right conclusion for the wrong reason, or brings to bear a seemingly unrelated body of knowledge that just happens to hold the clue to the right solution. . . . Consequently, market mechanisms and incentives only explain part of the story."

Either conception leads to the same conclusion, that each type of invention will have a different history and trajectory through the innovation cycle. Couple this with the distinction between potential adopters and a guaranteed base for a new technology and one can see why different factors in the fourth stage will become the dominant variables that explain the diffusion/adoption rate.

The work of previous researches gives primacy of place to one aspect of the diffusion process, but each qualifies that primacy by acknowledging on some level, the effects of other elements of the social structure. In the innovation model used in this book, these various elements are encapsulated and specifically set out.

Consequences

The consequences of a successfully diffused and adopted technology are twofold. The technology at once becomes part of the diffusion process in as much as it becomes part of the accumulated technical knowledge of the culture thereby giving it a role as part of the antecedent technology variable of the innovation process. Additionally, it can generate new needs in terms of improvements in the given technology as well as in terms of entirely new needs. I refrain from using the idea that a technology meets the need it was filling as that would imply that the innovation process was discrete, and as stated earlier, I believe the process to be continuous or perpetual. The history of human society points to the fact that we have perpetual needs grounded in the desire to increase control and human agency. Hence, rather than depicting innovation as a linear process, it is depicted as a continuous process. Figure 1-1 is a graphic representation of the modified model of technology innovation.

Figure 1-1. Modified Model of Technology Innovation

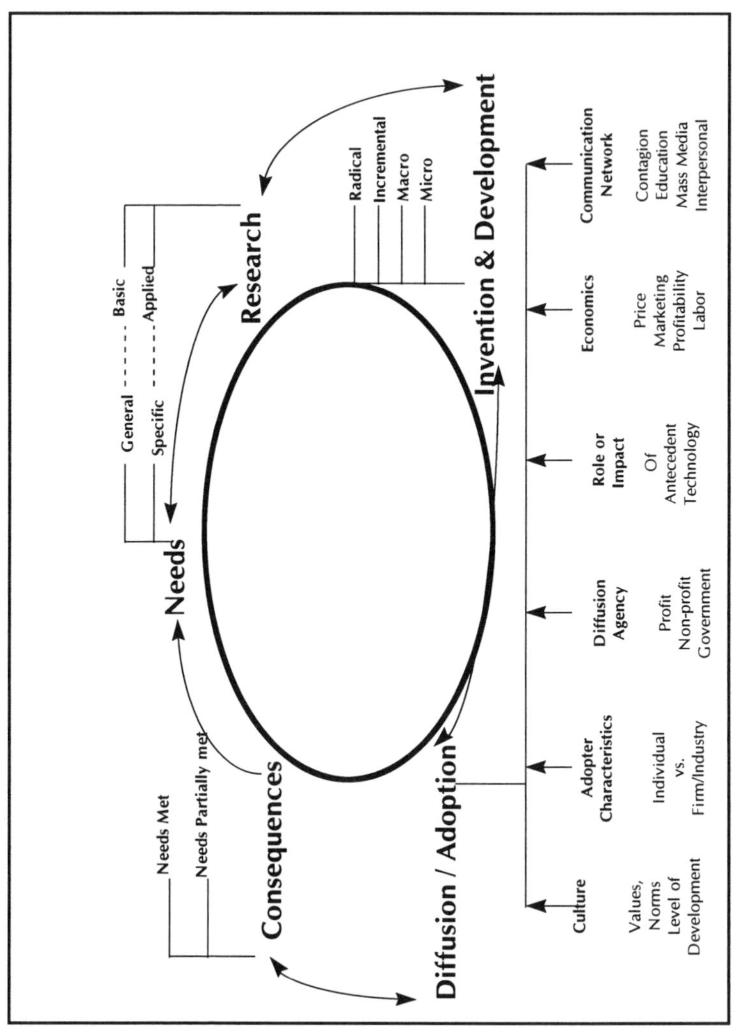

Diffusion of Interpersonal Communication Technology

The model for the diffusion of interpersonal communication technology posited below is meant to provide a generic pattern for such technologies. The exact trajectory of a new technology will depend on the kinds of factors that influence the direction and pace of events related to that technology. Consequently, the model is oriented towards the tasks/issues that need to be resolved in each stage. The cumulative resolutions will determine the trajectory of diffusion for each innovation. It is hypothesized that similar resolutions will yield similar results thus allowing for greater predictability.

Five stages are posited in the diffusion of an invention in interpersonal communication technology: ownership, development, infrastructure building, expansion, decline.

Ownership

Communication is the lifeblood of societies. The way we choose to interact is an integral element of the way we structure our society, and an essential element in affecting control. Because communication technology is a social phenomenon intimately related to issues of control, the question of ownership of a radical, new technology becomes an issue. The major choice is between government ownership and control, and private ownership and control. Each will have a different impact on the diffusion process. The issue of ownership is most crucial for diffusion after the initial invention, though it should be noted that questions of ownership and control do not always arise during the initial phases of diffusion, but will arise at some point.

Development

If the invention is a radical one and is to be developed by private industry, mass markets must be identified. Culture will play a major role at this point. The sense of need has to be created and cultural values and norms are used to awaken the collectivity to the necessity, desirability, and practicality of this new technology. On the level of the collectivity, or in other words on a mass scale, people need to be convinced that this new form of interpersonal communication technology is better than that which currently exists. Collectivities, as represented by organizations, industries, government or the public need to be convinced that this new technology will afford them greater control over their lives/work. Sometimes it will be the confluence of invention and historical condition that work together to aid in finding an initial market and application for the new technology. These initial markets will also be crucial in the development of the new industry that emerges as the result of the radical invention.

If the new invention represents an incremental change in technology, then a collective market is already defined, and the task is to display the superior qualities, applications and desirability of adopting the new technology for the collectivity.

If the invention is to be owned by the government, then a mass market is already identified and development depends on the governments ability/choice to find specific applications of the technology. Political ideology and applications assume prominence.

Infrastructure Building

If the technology was a radical invention then the infrastructure for that technology needs to be built. Typically, this stage will witness the introduction of many competing firms, each with its own variation of the invention and control over its portion of the physical network, and each incompatible with the others. Due to the systemic requirements of interpersonal communication, the ability of the new invention to perform efficiently and up to its potential will be hampered by the existence of such competing firms and equipment. The diffusion of the new technology will be slow or retarded due to the lack of standardization. Consequently, it will not be a period a large economic growth for the emerging industry.

Driven by economic pressures a natural monopoly will emerge. A particular company is likely to become dominant in the industry, and set the technical standards for the new technology. Additionally, the organizational form that has emerged will also be the standard for this type of technology. Once the infrastructure is built, diffusion can accelerate.

If the new technology was a singleton, or government owned, a single infrastructure will be developed and controlled by the dominant party, and it becomes both the technical and social standard for the new technology.

If the new technology is an incremental invention, it will be adapted to the infrastructure that is already in place as its vehicle of diffusion. Creating a technical standard for the new invention assumes primary importance in such a case.

Expansion

This is the period in which the new technology, if it is radical in nature, operates at its optimum level due to the benefits accrued from systemization, standardization and economies of scale. If the new technology is an incremental invention, growth is due in large part, to technical standardization. The role of diffusion agencies assumes great importance during this period as they represent systemic markets to which the new technology can be applied, and therefore significantly aid in the diffusion process. During this phase, the new technology becomes the social standard for the culture.

Decline

As the new invention becomes embedded in the culture, the initial period of diffusion will begin to slow and the decline can be the final one if the particular form of interpersonal communication technology is replaced by another radical interpersonal invention. Alternatively, if the technology continues to exist over time, the succeeding rises and falls in the diffusion curve over time may be explained by: a) Incremental improvements in the original technology, which are themselves subject to the particular political, cultural, and economic conditions of their historical time. b) A disturbance in the diffusion process due to a cataclysmic event such as war or economic depression. c) It could represent a partially unsuccessful diffusion and the subsequent industry adjustments made in marketing specific applications.

This model (graphically represented in Table 1-8) implicitly acknowledges the dynamic interplay of different variables which can impact on the diffusion process. It also accounts for historicity and points to the fact that over time, different variables assume primacy during the diffusion process. Consequently, a purely economic, social, political, or historical point of view in a model is inadequate to explain the totality of the diffusion process over time.

Table 1-8. Diffusion Model for Interpersonal Communication Technology

Ownership	Government-acts as the gatekeeper for technology diffusion into the private sphere and for the public at large. Private-economic institutions control and stamp the direction and diffusion of the new technology.
Development	If government owned, mass market is already identified and the focus is on finding specific applications. If privately owned, identifying potential adopters and convincing collectivities of the desirability of adopting and thus, beginning the diffusion process.
Infrastructure Building	If the invention is radical and privately owned, competing companies will emerge slowing the initial diffusion due to lack of standardization and a systemic network. Effort is directed at building the infrastructure required for successful diffusion. A monopolistic situation will emerge. The standards of the dominant company will become the technical standards for the industry. If the invention is incremental, it will make use of the pre-existing infrastructure, and, the quest for technical standardization assumes primacy. If the invention is a singleton or government owned, then a single infrastructure developed and controlled by that dominant party will become the technical and social standard for the new technology.
Expansion	For a radical invention, period of rapid growth due the benefits accrued from the development of the infrastructure and economies of scale. For and incremental invention, rapid growth is primarily due to standardization. In either case, there is a focus on diffusion agency as crucial to continued growth. Social standard for the new technology is set.
Decline	Technology becomes embedded in the social structure. It is replaced by an entirely new technology. Alternatively, there could be the continuing series of s-curves which reflect: shift in the application of that technology by the industry which was the vehicle for the initial diffusion; incremental changes in the technology which themselves are subject to cultural and economic forces; vagaries of history.

The next chapter provides an illustration of the innovation process and this diffusion model for interpersonal communication technology using telegraph technology as a case study, and suggesting it as a prototype for other communication technologies, such as the telephone, fax, personal computers and the Internet.[11]

NOTES

1. The major source for this table was the work of G.R.M. Garratt "Telegraphy" in *A History of Technology: the Industrial Revolution 1750-1850.*

2. In 1857 a German scientist and pianist, Hermann Von Helmholtz, discovered that he could make piano strings vibrate by singing into the piano. He also found that the strings would vibrate if he switched an electromagnet on and off, attracting the arms of a tuning fork, thus producing sound. In 1874 Alexander Graham Bell saw a phonautograph at work for the first time at the Massachusetts Institute of Technology. He learned the theory behind this from his contacts at MIT, and, combining this with his knowledge of the works of Oersted, Faraday, Sturgeon, and Helmholtz, Bell synthesized the knowledge and devised a system in which the sound that caused the first membrane to vibrate was reproduced by the second membrane. His new sound-reproducing machine eventually evolved into the telephone. In 1877, Thomas Edison, who had once been a telegraph operator, invented the repeating telegraph, or phonograph similar to Bell's device.

3. It is interesting to note that Garratt (1967) also tells us that first visual telegraph system in the United States was established for commercial use in 1800 by Jonathan Grout. His line connected Martha's Vineyard with Boston and was used to transmit news about shipping.

4. That people involved with telegraphy grasped the importance of the wedding of science to technology can be seen in the appearance of an electrician among the officials of the company in 1873 and of a research department within Western Union by the 1900's, as well as by the creation of schools dedicated to engineering endowed by prominent men in telegraphy, e.g. Ezra Cornell.

5. Two examples of engineering departments established by men interested in telegraphy are those at Cornell University, established in 1865, and Cooper Union, established in 1859. Ezra Cornell was involved in the laying of the first telegraph line and later went on to acquire the exclusive

rights to extend the use of the Hause printing telegraph system (it was the first to print Roman letters, numbers, and punctuation instead of the dot-and-dash code) throughout the United States, and which was the great-grandfather of he modern printing telegraph. Peter Cooper, most notable for his instrumental role in the laying of the transatlantic cable, established The Cooper Union for the Advancement of Science and Art. and among its course offerings were those in the technology of telegraphy.

6. Rogers (1983,p 12) defines innovation as "an idea, practice, or object that is perceived as new by an individual or other unit of adoption." Technology is a major component of the innovation process in that it allows for the reduction of uncertainty in the world, and for greater control.

7. An S-curve is a graphic representation of a performance parameter of a technology over time. According to Betz (1993) it does not explain the rate of change of a technology's progress, but rather, is an analogy of a commonly found pattern of incremental progress in a technology parameter over time. It is the graphic representation of choice used in the explanation of diffusion in this text.

8. Diffusion agency refers to the public or private sector through which an innovation is made available to the public at large. It will determine the of point of distribution as well as determine the level of access. Diffusion agencies establish infrastructures and operating procedures to promote adoption. Incorporated into this perspective is an acknowledgment that the infrastructure that already exists in society interacts with that developed specifically to deal with innovation diffusion to constrain or enable the diffusion process and together form the supply and demand dynamics that are the focus of this perspective.

9. Universal access is defined as the "ability to reach all members through a medium" and "is an important outcome for any community into which an interactive medium is newly introduced (Markus 1987: 491-492)

10. Oliver, et al (1985: 524) explain critical mass as "a small group of the population that chooses to make a big contribution to the collective action, while the majority do little or nothing".

11. It is important to note that it was the standardization of the communication protocol that allowed for the diffusion of Internet, which is actually a collection of networks that are interconnected by this common communication protocol, or language.

II

Telegraph Innovation: Origin and Diffusion

This chapter discusses technology innovation and applies the author's diffusion model for interpersonal communication technology to telegraphy in order to illustrate the total innovation process. S-curves representing telegraph and telephone diffusion are presented and analyzed in light of this model.

ORIGIN

Basic research is a continuous process in the quest of human beings to manipulate and control their environment. Betz (1993: 129) tells us that "Invention is the creative process in which new logical ways are imagined to manipulate nature for human purposes." Telegraph technology was invented as a response to a general need for faster communication. This also meant that telegraph technology had only an amorphous pool of potential adopters. The initial adopters of this new technology would be those who would first be able to discern a practical application of this new technology as it had no reference point within American culture at the time. The idea of instantaneous communication was new to the culture and ways had yet to be devised for incorporating it into the social fabric or ways of being in society. Looked at from another way, one could say that specific needs had to be imagined and cultivated. If one looks back to the table depicting the early history of telegraph technology (Table 1-1) one can also see the influence of previous research and technologies. The telegraph would not have been possible without the accumulated cultural knowledge and without the interplay between science (basic research) in terms of the principles of electricity, and applied research, that is, applying this knowledge to solving the need for greater control over communication at a distance. Samuel F.B. Morse, responding to a sense of general need and driven by the findings of research, was able to apply the principles of electricity to the service of communication, and in 1838 Morse invented the telegraph. This represented a radical innovation in that it gave human agency to the control of information processing, mediating the effects of weather and distance for the first time in human history. Telegraphic technology made possible instantaneous, point-to-point communication, also a first in human history. Morse, an independent inventor (who incidentally was an artist by trade), needed funds to bring his invention into a form in which it was practical to present to the public for adoption and diffusion. He sought funding from the federal government. On February 21, 1838, President Martin Van Buren and his Cabinet witnessed a demonstration of Samuel F. B. Morse's electric telegraph. Amazement! April 6, 1838, Congressman Francis O. J. Smith submitted a report to the House of Representatives in which he writes: "It is obvious, however, that the influence of this invention over the political, commercial and social relations of the people of this widely extended country . . . will amount to a revolution unsurpassed in moral grandeur by any discovery that has been made in the arts and sciences . . . " Morse (1914: 87)

However, amazement did not immediately lead to action. It wasn't until March 3, 1843 that Congress appropriated $30,000.00 for Morse to experiment with his electro-magnetic telegraph. The first telegraph line in the United States was constructed between Washington, D.C. and Baltimore, Maryland. It was completed on April 28, 1844, and, mirabile dictu, it worked! May, 1844, the formal use of the Morse telegraph line began with the words "What hath God wrought!" It was considered a radical invention because it was the first modern communication technology that had the ability to significantly alter interpersonal communication relations; it gave increased human agency to long-distance communication. One can get a sense of the magnitude and "radicalness" of this new technology when twenty seven years after its initial introduction, at the dedication ceremony of a statue in honor of Morse, the telegraph was still referred to with wonderment:

> "In our day a new era has dawned. Again for the second time in the history of the world, the power of language is increased by human agency. Thanks to Samuel F. B. Morse men speak to one another now, though separated by the width of the earth, with lightning's speed and as if standing face to face. If the inventor of the alphabet be deserving of the highest honors, so is he whose great achievement marks this epoch in the history of language—the inventor of the Electric Telegraph."

John T. Hoffman
Governor of New York
June 10, 1871

DIFFUSION

Ownership and Development

Morse proved the telegraph proved to be a viable invention. However, as this invention was radical in character and the result of general need, there was no ready pool of adopters waiting to adopt Morse's invention and begin the diffusion process. Morse looked to the federal government to assume ownership of his invention. Morse wanted the United States government to purchase all rights and patents to the telegraph so that it would be integrated with the postal system as it later was in much of Europe. Morse conceived of telegraphy as belonging in the political sphere as optical telegraphy was in Europe, where, because of almost constant warfare, control over communication and information was a primary concern of the governments. Hence, Morse, knowledgeable about European affairs as well as having traveled in Europe, expected the government to adopt this new technology for their own needs and ends and incorporate it into a government agency. This was not to be. The telegraph industry in the United States emerged as a purely commercial enterprise. Coinciding with the rise of the ideology of laissez-faire capitalism, the U.S. government did not want to get involved in the development of telegraphic technology beyond its initial appropriation of $30,000.00. In opposing, in 1845, the bill for the extension of the government line, one of the members of Congress, Thomas H. Benton, reflected the feelings of the government when he said that he had rejoiced at the invention of the telegraph and hoped to see it extended to all principal cities of the country, but that he "wanted it to be called for by the commerce of the country and to pay its own expenses."[1] According to Marshall (1951) the telegraph line made $413.00 and lost $3,274.00 in its first 6 months of operation. Clearly, the government didn't think this would be a profitable enterprise. Consequently, when it became clear in early 1845 that the government would not take control of his invention, Morse set about finding the best means of developing telegraphy through private enterprise. Morse chose Amos Kendall, a lawyer and former Postmaster General under President Jackson, as his business and legal advisor. Kendall was able to induce a number of entrepreneurs into subscribing to what in 1845 was still a risky venture. Kendall soon formed the first telegraph company in the United States—The Magnetic Telegraph Company—with Morse as the major stockholder. In May, 1845, the company made plans to build a line from New York to Washington, D.C., the first step in constructing a vast system that would ultimately wire the world.

It should be noted that this was only the first issue of ownership that was to arise regarding the telegraph. However, it can be considered the most crucial for the development of telegraphy because it stamped the major direction for that technology, namely that it was for commercial use and therefore theoretically accessible to the entire population. This is in contrast to France for example, where the telegraph was owned and controlled by the government. It took five years for the government to grant access to the public, and even then, government dispatches still took precedent. Chapter 4 deals with the issue of the ownership of telegraph technology as it arose again in 1866 and during World War I.

The invention of telegraphy spawned the birth of a new industry. Consequently, when discussing the diffusion/adoption of telegraph technology, we must separate it into two aspects: growth of the industry itself; diffusion into the general social structure (that is, its adoption by the military, other business firms, and the public in general). The newspaper industry proved to be the first collective adopter of the telegraph and critical to spurring the initial building of telegraph lines. This was a result of the convergence of an historical event, The Mexican-American War, and need, as the people of the United States wanted news of the war as fast as possible. Thus, it is important to discuss the relationship between telegraphy and the newspaper industry when chronicling early diffusion and development.

The Newspaper Industry

The newspaper industry was among the first adopters to find a use for this new telegraph technology, and that need came to the attention of the newspaper people via the outbreak of the Mexican War. Thompson (1947) indicates that the outbreak of the Mexican War in 1846 virtually forced the adoption of the telegraph by the newspaper press. U.S. citizens had an insatiable demand for news from the front. Journalists were challenged to find more rapid means of forwarding the news of the battles. Telegraphic technology was there to answer the call. Alexander Jones, a pioneer in the field of telegraph reporting wrote " . . . the utility of the telegraph to the press in forwarding army news was such as, in a measure, to force them into its employment." (Thompson 1947: 220).

The existing, competing newspapers in the Northeast were already spending lots of money on pony expresses, special trains, rowboats and clipper ships to bring them news. The Mexican War meant that, as Thompson (1947: 217) tells us, " . . . the prospect of even greater distances to cover caused concern . . . " Bennett and other news publishers made what use they could of the fragmentary lines in operation at the outbreak of the Mexican War (that is, the experimental line between Baltimore and Washington, D.C., and another one that was under construction between Philadelphia and Baltimore, and the one between New York and Washington, D.C. that was not complete until 1847) and those that came into being during the war. Thompson (1947: 219) tells us that "A week after the outbreak of the war, dispatches were being sent from Washington to Baltimore, and thence by special steam and horse express to Wilmington, from which point they were telegraphed to Jersey City" (via the Magnetic Telegraph Company line that was under construction).

The attention of the press on the telegraph industry and its ability to aid in timely newsgathering about the war, encouraged private investment in the budding telegraph industry. In one sense, then, the Newspaper industry can be thought of as a diffusion agency as well as an initial adopter. The adoption of the telegraph during the Mexican War aided the diffusion process in two ways: creating a base of adopters in that all newspapers that wanted to remain viable had to adopt telegraph technology as it quickly became the industry standard; it created an additional impetus for expediting the construction of telegraph lines as the newspaper industry directly benefited from the increased physical diffusion of this technology. Thompson (1947: 219) informs us that "Telegraph lines in the blueprint stage in 1846 had become reality by the summer of 1847." As the nascent telegraph industry continued to extend its coverage over the land, the press came to depend more fully on it for transmitting the news. The telegraph mediated the experience of news gathering by mitigating time-space distantiation. This implies that the interval between the occurrence of a newsworthy event and its appearance in print was almost instantaneous.

Infrastructure Building

Shortly after the New York—Washington, D.C. line was completed in 1846, other companies began business. Some were licensed by the owners of Samuel Morsepatents, and others were based on rival technologies. For example some companies used the printing telegraph based on Royal E. Hause's invention whereby messages were printed on tape or paper in Roman letters (as opposed to Morse's dot and dash system). In 1848, Alexander Bain received patents on yet another variation of the printing telegraph. These were but two of the many competing and incompatible technologies that developed. This cornucopia of independent telegraph companies combined with the use of various technologies resulted in confusion, inefficiency and a rash of suits and countersuits concerning patent licensing and infringement. By 1851 there were over 50 separate telegraph companies operating in the United States. Communication, though instantaneous, was, in actuality, only relatively so. Often a message had to go through a number of independent companies before it reached its destination. There was no uniformity of tariff (making telegraphing expensive in that it often had to be paid for through several companies) and no uniform processing procedures or delivery procedures. Often messages were mis-translated or lost within the maze of multiple transmissions. Operating as independent competitors, firms often became unprofitable. It became clear that the segmented market was insufficient to the needs and scope of telegraphic communication. While the isolated companies may have been able to meet local needs, the telegraph was an instrument that had *systematic* needs. In other words, there had to be a smooth flow of information over diverse geographic areas that were interconnected in an uninterrupted fashion. It was necessary that each facet of the operation interfaced and cooperated harmoniously with each of the others. And the more complete the network, the more useful it would be.

It was the foresight of a few businessmen that turned a floundering telegraph industry into a telegraph system. The birth and subsequent history of Western Union is testament to this fact. "While the majority of small companies were limping into bankruptcy, a group of Rochester, New York business men, headed by Hiram Sibley, realized the salvation of the telegraph industry depended upon unification. They formed the New York and Mississippi Valley Telegraph Company, gathering into it a number of feeble companies. Together they made a strong bundle— strengthened by purchase of patents and additional lines—reorganized in 1856 by Ezra Cornell, who was the largest stockholder, and called Western Union." (White 1939: 10)

The Telegraph and the Railroads

In its infancy, the growth of telegraphy was intimately tied to the growth of another industry (also in its infancy), the railroads. The role of this almost concomitant (as opposed to antecedent) technology in this aspect of telegraph diffusion is indispensable and inseparable. It is impossible to discuss the evolution of telegraphy as an industry without first discussing its relationship to railroad technology. They were complementary technologies; each meeting particular needs of the other, and in the case of the telegraph, enabling it to become a monopoly.

Hughes (1989:4) tells us "Inventors, industrial scientists, engineers, and system builders have been the makers of modern America. The values of order, system, and control that they embedded in machines, devices, processes, and systems have become the values of modern technological culture." Communication, rapid and efficient, was essential to their efforts. There were a number of factors which dictated the direction of much of this inventive effort. Among the most important according to Hughes (1989) were: the vast area which comprised the United States; the needs of commerce; the needs of politics. All of these, Hughes (1989: 4) states "encouraged the invention of a set of machines designed, in one way or another, to facilitate communication."

Hughes (1989: 3) also states that "The electric telegraph . . . was . . . the first large-scale and commercially important use of electricity." The railroad companies were the prime users of telegraphy. These two industries developed hand and hand, spreading a vast transportation and communication network over the entire country. Chandler (1977: 195) tells us "The railroad and the telegraph marched across the continent in unison." Telegraphy enabled the railroad industry to grow by facilitating efficient management, as well as increased safety and speed. Railroads enabled the telegraph industry to grow by allowing the erection of telegraph polls alongside railroad tracks on land for which the railroads had already gained the right of way. This joint effort of technological growth and dissemination was accomplished through mutually beneficial contracts between the railroads and the telegraph companies. A brief history of the early development of the railroad and telegraph industries will demonstrate the shifts in power and control that the telegraph facilitated. It will also make clear the way the telegraph became an integral and invaluable tool of business.

Chandler (1977) provides an insightful history of the railroads as the first modern business enterprise developed between the 1850's and 1860's, and much of what follows is a summary of his work as it relates to the relationship between the railroads and telegraphy.

The first railroads in the United States were constructed on a single line of track. Head-on collisions between locomotives were a major hazard. This meant that the movement of trains had to be carefully controlled if goods and passengers were to be moved with safety and a degree of efficiency. It was a technology that had systematic needs in order to effect control. The early optical telegraph (semaphore) was able to provide a modicum of control, but was limited by the vagaries of weather and the time of day. The railroad building boom that began in the 1840's only intensified the need for more efficient communication in the railroad industry. The invention of the telegraph in 1838 and its subsequent adoption for commercial use in 1847, provided an answer to the control needs of the railroads. Chandler (1977,: 89) indicates "Railroad managers quickly found the telegraph an invaluable aid in assuring the safe and efficient operation of trains; and telegraph promoters realized that the railroads provided the only convenient rights of way". Thus, we see the basis for mutual cooperation and development.

As previously mentioned, the early history of the telegraph industry shows that initially, a number of independent telegraph companies emerged.[2] Rather than negotiating for their own rights-of-way, these companies allied themselves with the railroad early on and, as Chandler tells us, some were actually subsidiaries of the railroad companies. However, the telegraph companies that were subsidiaries of railroad companies faced the same problem that the independent companies faced—the systemic need of the telegraph. Consequently, the railroad telegraph companies succumbed to the pressure of Western Union's system building and were either bought out by Western Union or forced out of business. Figure 2-1 depicts the extent of Western Union's consolidation efforts in building the telegraph system.

The marriage of these two industries, as Reid (1879: 480) indicates, "went on until the whole railroad system of the continent has become more or less identified with telegraph interest, and with its whole transportation service regulated by its control." A sample summary of early contracts between the railroad and telegraph companies shows the mutual benefit each of these industries received. (See Figure 2-2) Reid (1879: 480) also tells us that it is difficult to ascertain which industry benefited the most. He says:

> "To the railroad companies it is of inestimable value. Not only does it give to them an economical means of control, and a share of telegraphic revenues, but the executive officers are usually

provided with such wide facilities for communication with other companies, and with their own officers at distant places, that the whole machinery of management has become sublimely prompt, consolidated and simple. To the telegraph company it is protection, economy, permanence and strength."

Figure 2-1. Diagram of Western Union' s Consolidation Efforts Through 1915

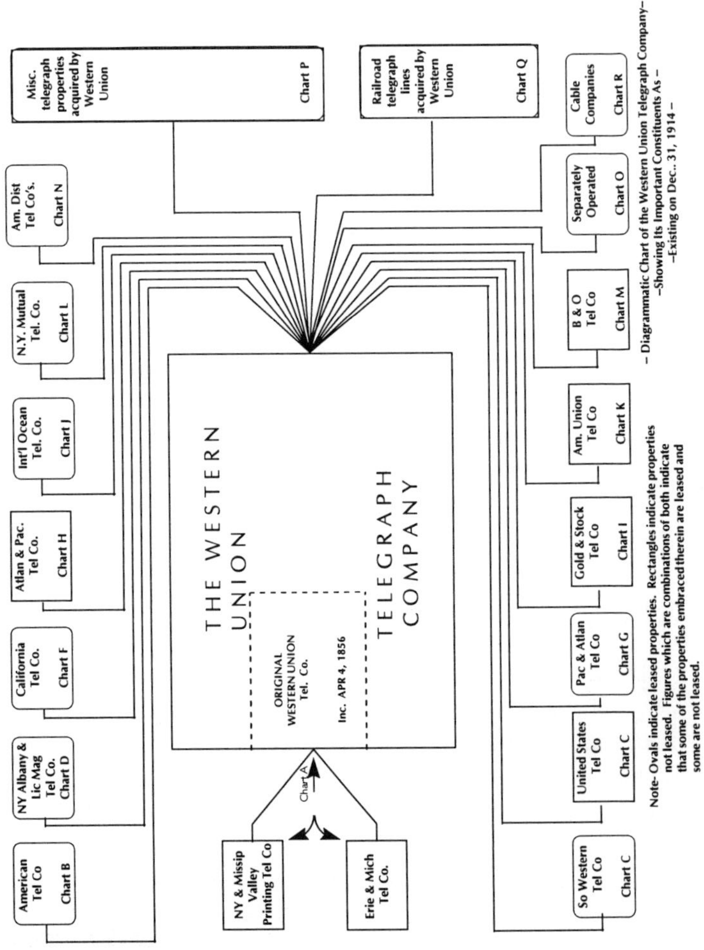

Figure 2-2. Sample Summary of an early contract between a telegraph company an a railroad company. (From Reid (1879: 480-481)

The telegraph company to furnish the railroad company with a single wire of proper size and quality, and provide Morse instruments at certain specified stations on the line of their road.

The telegraph company to maintain main battery for working said wire day and night.

The telegraph company to keep the wire erected for the railroad company in order, except as otherwise provided.

All receipts for messages at offices opened on the line of the railroad, by either party, to belong to the telegraph company.

The railroad company not to send any message free except for its own agents on its own business.

At all stations in addition to those named, the railroad company to supply all machinery and local battery.

The railroad company to instruct its men to watch the line, straighten poles, re-set the same when down, mend wires, and report to the telegraph company.

The railroad company to convey and distribute wire and insulators and all other material free, and also furnish a hand-car for stringing wire.

The two companies to reciprocate the use of wires when those of either are out of order, but railroad wire never to be interrupted when sending railroad business.

The railroad company to transport all instruments, material for repairs, all operators, officer and agents of the telegraph company free of charge when on business of the company, and to furnish and distribute poles when line has to be renewed, the telegraph company setting and insulating the same.

The railroad company to pay for stringing the railroad wire and insulating the same, and for instruments, etc., thirty dollars per mile.

The railroad company not to allow any other telegraph company to build a telegraph line upon its property.

Railroad telegraph operators may accept public business at the ordinary tariffs, and shall account for the same to the telegraph company, but no messages will be accepted or sent to interfere with railroad business.

Item number 12 "The railroad company not to allow any other telegraph company to build a telegraph line upon its property." became an invaluable clause in the evolution of the telegraph industry by effectively providing a means to limit competition along a given line. The telegraph company who had wires along the right-of-way of the railroad benefited from economies of building and operation not available to competitors. Consequently, they were often able to charge cheaper rates, and thereby drive out or annul any possible threat from a competitor. The systemic needs of the telegraph industry, aided by such agreements, facilitated the growth of the first modern business monopoly, Western Union.

The physical infrastructure that was built by the telegraph industry was crucial to its successful diffusion, and monopolization was crucial to maintaining control over that infrastructure in order to maintain the smooth, continuous, rapid and uninterrupted flow of messages across the wires. Agreements between several big firms to standardize equipment, message forms or operating procedures would not have been sufficient for successful diffusion because it would have meant decentralized control over the physical infrastructure. If each company owned their own wires the systemic needs of the technology would have potentially been at risk at any time in relation to the relative stability/instability of a given firm. What would have happened if a firm was not maintaining its wires properly, or if they were experiencing financial difficulties, or went bankrupt? Consolidation would likely have taken place, and in fact did take place, in the course of the system building that Western Union accomplished. Centralized and unitary control, or monopolization, was essential in the successful diffusion of telegraph technology.[3]

The fusion of railroad and telegraph technology allowed for the emergence of what Thompson (1947: 216) called "strong and aggressive communication and transportation systems which transformed the life of the nation." However, Chandler (1977: 147) tells us that real system building, that is, creating a structure which assured a "continuing flow of freight and passengers across the roads' facilities by fully controlling connections with major sources of traffic", did not occur in the railroad industry until the 1880's, by which time the railroads had been integrated into a single national network. Consequently, the control made possible by the rapid communication that the telegraph provided, did not achieve its fullest effect until the late 1800's when the other economic vagaries were dealt with (for example, effects of competing companies, managerial issues and the impact of small, entrepreneurial enterprises—which effectively disappeared by 1900).

Expansion

During this phase in the diffusion process the role of diffusion agencies becomes crucial. The United States government, via its political policies and via the military, and the business community via inter-industry linkages were primary diffusion agencies for the new systemic telegraphic technology. In addition, it should be noted that the railroads and the newspapers continued to be important diffusion agencies throughout the history of the telegraph, though their most crucial impact was during the telegraph's budding infancy as they directly aided in initial development as discussed in the previous section.

The government played a big role in the diffusion of telegraph technology. Noble (1977) suggests that it is often the government that is important in the development and dissemination of a new technology. Though the government was not the primary underwriter of telegraph industry development, it was an important diffusion agency. The main infrastructure that the government used to diffuse this new technology was the military. The military facilitated the diffusion of telegraphy in three major ways: as the result of war; as part of its policy to aid in the commerce of the nation; as part of its political policy of expansionism.

The Telegraph and the American Civil War

Clearly, control over information was advantageous in wartime. Plum (1882: 10) writes "It is believed that no nation was ever, in times of war, content to await even the speediest methods of conveying news of battles in which its forces were engaged . . . All wars illustrate the importance of speedy communication." Just as the networking value and capacity of the telegraph was being experienced by the railroad industry, the American Civil War brought to public attention the value of the telegraph in an emergency, where timing and control of information were crucial. War dramatically points out the vital role of communication technology for the dominion of human beings over allocative (material) and authoritative (social) resources of the world.

Rupp[4] (1990: 1) informs us that "The War of the Rebellion, as the Civil War was officially known, was the first war in which technology played such a dominant role." According to Rupp, the technological innovations that were most important in terms of military operations were: the use of ironclad ships—which "set the pattern for all naval vessels to follow"; railroads—which significantly altered the way armies moved and supplied themselves; and the telegraph—which altered the logistics of war.

Up until the development of the electrical telegraph there were no dramatic changes in the way generals communicated with commanders. Rupp (1990: 2) writes "During the Revolutionary War, General Washington had no better means of communicating with his commanders than did his counterparts of the previous 2,000 or 3,000 years." Furthermore Rupp indicates "Improved communication was always a matter of importance to the army. Consequently, when the technology became available to enhance communication via signaling, the army adopted it. The first system the army adopted was essentially a semaphore system, one enhanced and standardized by First Lieutenant Albert J. Meyers, in 1858. According to Rupp (1990: 2) "It proved superior to anything then in use." By 1860, the position of Chief Signal Officer was added to the army's General Staff. One year later, with the outbreak of the Civil War, an "Acting Signal Corps of the Army" was formed. By 1863 Congress officially recognized this unit, calling for it to remain in service until the "rebellion" was over. The war caused Chief Signal Officer Meyers to recognize the limitations of visual telegraphy and initiate the use of electrical telegraphy by "building short field lines that tied various headquarters together, and then extending them to connect with railroad or commercial telegraph lines."

The introduction of the telegraph, in concert with the development of the railroads and improvements in weaponry, drastically altered the ability for control. The telegraph did much to centralize the decision-making process during the war by making possible rapid communication between the War Dept., the President, and the different divisions of the army. Planning became more efficient as one of the vagaries of war—"intelligence" or information gained human agency. Prior to telegraphy, orders had to be sent via people. This was often an inefficient means in that people could be intercepted en route to/from message delivery, taken as prisoners of war, or killed. Control over information was essentially beyond human control so that often by the time information reached its destination it was outdated and troops may have moved. Telegraphy changed all that inasmuch as it allowed for up to the moment information processing which greatly aided planning and control. The use of the telegraph during the Civil War caused shifts in power at many levels, and enhanced the ability to control at all levels. The introduction of mobile field telegraphs made possible centralized control of field tactics thus enhancing the generals' ability to control troop movement and maneuvers. It enabled a general to keep in touch with several corps at the same time and coordinate their activity, all from his headquarters. Additionally, the generals were able to keep in constant contact through the general military telegraph system, with each other and the War Department. The Civil War was the first in which generals were able to direct maneuvers in response to actual, rather than anticipated movements or outdated information. For example, telegrams advised generals of each other movements. An August 19, 1862 telegram sent to General Parker advised: "The remainder of Seymour's and Jackson's brigades will be here this afternoon, . . . " (*War Telegrams* Thomas Dolan Collection 1861-1863). Or, another example, also from the Dolan Collection, reflects the ability for increasing logistical control: "I command in the extreme left and if attacked will have to repel with a limited supply of ammunition . . . Please inform me when a supply can be sent." It was clear that the telegraph was of prime strategic necessity. The telegraph facilitated the execution of battle plans made at the level of the commander-in-chief. A more controlled application of these plans was made possible, as well as the necessary up-to-the-moment revisions, due to telegraphy. Rupp (1990) reports that President Lincoln could be found most evenings next to the telegraph. In fact he states that "For the last two or three weeks of his life Lincoln virtually lived in the telegraph office . . . and the wires were kept busy with dispatches to and from the President." No wonder that telegraphers were among the first prisoners of war to be taken. Coates, et al (1979). Disrupting communication was of prime tactical importance.

The Union army is hypothesized to have had an advantage during the Civil War due to its superior telegraph system. The North had a vast communication network manned by expert telegraphers, as well as cipher specialists.[5] The government was able to enlist the services of the most prominent men in telegraphy to serve as leaders of the military telegraph. Anson Stager, who was made a general, was called by General McClellan to organize the military telegraph. Reid (1879) tells us Stager was good at cryptographs and well versed in running a telegraph company. It was not uncommon for encoded messages traversing several telegraph lines to be intercepted and decoded by these Northern cipher specialists. O'Brien (1910) posits that the less advanced telegraphy in the South did not produce such talents. He asserts that the faithful service of Southern commercial and railroad telegraphers was no substitute in the Confederacy for a well organized military telegraph. O'Brien suggests that the use of the telegraph in battle can be the difference between victory and defeat—going as far as to suggest that the well-developed military telegraph system of the North precipitated its victory.

> "It appears that the want of coordination in the movement of Lee's columns, and the ignorance of the whereabouts of his cavalry under Steward, precipitated the Battle of Gettysburg at a point he would not have chosen, and that in the actual climax of the battle, the wings of his army attacked separately and not together. Therefore, it seems logical to suppose that if Lee had had a military telegraph such as ours, or like that by which the Germans afterwards moved their separate columns in the Franco-Prussian War of 1870, he might have been able to avoid Gettysburg or select a more favorable ground for the battle, and, in the actual fight, to have attacked with all his forces synchronously."
>
> O'Brien (1910: 295-6)

The Civil War allows us to see how a more effective means of information processing, coupled with its amenability to networking, allowed for the centralization of power and control, which in turn significantly altered the logistics of war.

The Telegraph and the Governments Policy to Aid Commerce: the Coast Guard and the National Weather Service

The relationship between the telegraph and the military can be examined from another perspective. While the impact that telegraphy made on the military concerning an increase in its ability to control and the consequent enhancement of power was just discussed, the relationship was actually multi-faceted. The government, during the course of the Civil War, constructed 15,000 miles of telegraph lines (Coates 1979). This construction however, did not represent any appreciable permanent dissemination of telegraphic technology in that it was of inferior quality (expecting only to be temporary) and much of it was abandoned after the war. Rupp (1990: 2) tells us "The lines the army built during the war were all of a temporary nature and were soon abandoned, once the reason for their existence ceased. It was in the postwar era that the military became an important factor in telegraph history". After the Civil War, appropriations to the army were greatly reduced. Rupp (1990: 2) also informs us that "Working under these tight reductions, the main object [of the Signal Corps] was merely to maintain the knowledge obtained in military telegraphy during the war, as well as to provide a nucleus of trained personnel capable of expansion in an emergency." The continuation of the Signal Corps was tenuous at best. The first post war use of the Military Telegraph system was a modest effort whereby the forts within certain harbors were tied together internally." This statement gives us the first hint of a way in which the government indirectly contributed to the shaping of telegraphic history. One of the military's main contributions to the diffusion of telegraphic technology came in the form of a new application—national weather forecasting. A *Scientific American* Article dated Jan. 5, 1921 informs us Professor Henry of the Smithsonian Institution is credited with showing how the telegraph could be used in weather reporting. He felt that such forecasts would protect commerce by warning of storms, monitoring the levels of water in rivers, and noting temperatures, all of which affected shipping. In Feb., 1870, the U.S. Army Signal Corps received a new mission and its continued existence was guaranteed, and its part in the creation of two institutions was insured. Public Resolution No. 9, passed on Feb. 9, 1870, required the "Secretary of War to provide for the taking of meteorological observations and to give notice of approaching storms on the Great Lakes and along the seacoast." (Rupp 1990: 6) Through this resolution Congress required that the meteorological readings be telegraphed to the various signal stations along the coasts "for the benefit of the commerce of the United States." (Rupp 1990: 7) These signal stations were connected to the office of the Chief Signal Officer of the Signal Corps. Thus, the government effectively established the first national weather service. These forecasts were often picked up and printed in newspapers across the country as they were also of great benefit to farmers.

The military also established the "Life Saving Service" in 1871. It was the responsibility of this Service to coordinate the signaling service provided along the seacoast to warn ships of approaching storms. It also provided rescue aid to ships affected by disaster. This Service was created by the government primarily through the take-over of the privately run life saving stations, as well as through the construction of new ones. This "Life-Saving Service", initially operated by the Revenue Cutter Service, a branch of the Treasury Dept., became an independent bureau of the treasury in 1878. It eventually evolved into what we now know as the United States Coast Guard by 1915. (Rupp 1990).

Through the auspices of these two institutions the military extended telegraph service along the east coast, connecting it at various inland points with commercial telegraph lines. Rupp (1990: 11) informs us, "In 1883-84, Congress reduced the appropriations available to the Signal Corps for telegraph line repair and maintenance. As a result, on October 1, 1884, all of the leased lines connecting the seacoast lines with the office of the Chief Signal Officer were given up, and thereafter all communications were sent over the wires of Western Union." This represents the government's relinquishing of control over telegraphy in favor of private enterprise, which was nothing new, but points to one sound reason for it—economy.

The Telegraph and Government Expansionist Policy

Post—Civil War America was involved in establishing itself as a powerful entity in the global community. In order to accomplish this, an increased sense of nationalism had to be developed (see Chapter 8 for a full discussion of this concept) as the United States sought to create an image of itself as a "state"[6] in the world polity. During this post-war period the United States was interested in unifying the nation in terms of attitude and in repairing the psychological damage done to the sense of unity in this country due to the Civil War. Concomitantly, the government was also interested in clearly laying claim to and demarcating the boundaries that made up the United States of America. Consequently, the government was interested in expanding its boundaries and increasing its resources that this expansion made possible. Additionally, in positioning itself to be a political entity in the global community, geographic control was very important. According to Axline and Stegengal (1972) natural boundaries render a state less accessible to invasion, and, that a nation's resources are an essential element in the determination of national power. The telegraph mediated the experience of the expansion.

In order for the frontier to be settled, migrants needed to feel protected from hostile peoples. The Military Posts in the West served this purpose. Enhancing the sense of security and decreasing the sense of isolation was the function of the telegraph on the frontier. A few excerpts from the 1880 Annual Report of the Chief Signal Officer will illustrate just how vital these military telegraphs were to the public, the settlers.

> "The military telegraph line is of great benefit to those in the past and the scattered settlements to the north and east of us. Parties frequently come in from St. Johns and other places over a hundred miles from here to place themselves in communication with the telegraph stations of the world." Rupp (1990: 114)

> "The fact of speedy means of communication gives a feeling of security that would not prevail were telegraphic facilities removed, and the only regret of the settlers is that there is not more commercial business done here to recompense the government for

the great benefit the community derives from the line." Rupp (1990: 116)

"During the past year the United States military telegraph line has been of incalculable value to the public at large, and the military authorities in particular."[7]

One of the Military's missions, according to Rupp (1990: 17), in the post-civil war era, was "The construction of thousands of miles of telegraph lines . . . " under the direction of the Signal Corps of the army. The government via the military, built 7,355 miles of telegraph lines throughout the west and southwest, connecting these areas with the settled areas of the country between 1871-1881. These lines were built, operated and maintained by the army. And, as Rupp (1990: 24) states "The military telegraph system certainly wasn't as romantic as the Pony Express, yet in its own quiet way was far more influential in shaping the course of events, as it provided the only reliable means of communication for many places on the frontier."

These lines were not only of military importance (especially in dealing with the Indians) but had social and economic value as well. For example, when considering the construction of a telegraph line from the Black Hills to Forts Keogh, Custer and Ellis in Montana, one of the arguments for construction was the following: "Its maintenance will cost less than the usual military couriers sent from post to post, using up and killing annually many horses. This line will furnish prompt intelligence along a line which soon must be occupied by settlers and will greatly facilitate such settlement; soon to be followed by the usual train and wagon travel." Rupp (1990: 22) It was also assumed that after these areas were settled, the settlers would make much use of the lines so that eventually the lines would pay for themselves.

The Military Telegraph system reached its peak around 1880-1881. (Rupp 1990: 24) informs us that there were basically two reasons for the decline of the military telegraph system. First, "As the railroads and their associated telegraph lines expanded into the country, there was no reason for the military to maintain duplicate facilities, and thus the military lines were either sold or abandoned." Given the increasingly restrictive budget of the Signal Corps, it did not make economic sense for there to be a duplication of services. The army's policy of abandoning lines as soon as commercial lines could provide the necessary service, reduced the military system to 1,025 miles of wire by June, 1891.

Second, "the rapid settlement of the western territories reduced the need for the numerous far-flung army posts." Additionally, as the west and southwest became more densely populated, their needs were more in line with those that could be better satisfied by the railroad and commercial lines which were not as fragmented or outdated as the military lines. Again we see the government choosing not to compete with private enterprise in the telegraph industry.

It is very important to note that the government, during this period of telegraph diffusion, was active on two fronts. At the same time that it was using federal funds via the military to build a telegraph system, the government was also aiding the commercial development of telegraphy through land grants. Rupp (1990: 85) tells us "The first Pacific Railroad Act, passed July 1, 1862, granting lands and bonds in aid of 'a railroad and telegraph line from the Missouri River and the Pacific Ocean,' authorized and empowered the Union Pacific Railway Company to 'lay out, locate, construct, furnish, maintain and enjoy a continuous railroad and telegraph.'" Additionally, on July 24, 1866, an act of Congress was approved which effectively insured the primacy of government access to any lines it aided in the construction of either directly, through monetary grants, or indirectly, through land grants. Rupp (1990: 85) states "A number of courts had upheld these laws and the government's right to use these lines by noting: The grant of public lands and of public credit in aid of the construction of the railroad and telegraph lines, even without express provisions, imply a public use of them." Rupp further states "As just about every major railroad in the west and its associated telegraph line had received some sort of government aid, usually in the form of land grants, they were obligated to transmit government business over their lines." Thus, we see that there was much interrelatedness between the telegraph industry and railroad industry growth, and the government, in its role as primary facilitator.

Inter-industry Linkages: the Telegraph and Market Control

From a systemic point of view, it is clear that telegraph and railroad diffusion led to further diffusion of both of these technologies. The telegraph and the railroad, working in concert with each other, also exerted a great influence on the economic sphere of life via their ability to facilitate market control. More and more industries became dependent upon these two technologies, increasing the need for them and thus aiding in the continuing diffusion of both technologies.

Commerce is the area in which we can most clearly see the effects of the transformation Thompson (1947) refers to when he said that telegraph and railroad technology transformed the life of the nation. These two technologies increased businesses' ability to control, which led to greater efficiency, which in turn generally translated into greater profitability.

According to Chandler (1977: 37) "Prior to 1840, new technology had not yet lifted the age-old constraints on the speed a given amount of goods might be moved over a given distance." Additionally, Chandler (1977: 38) informs us that "Although merchants wrote long and detailed letters of instruction to correspondents, captains, or supercargoes, they had little control over the actions and decisions of their agents in distant ports or on distant seas. Letters took weeks or sometimes months to reach their destinations. Only the man on the spot knew how to adjust to changing local market conditions." The concurrent development and dissemination of railroad and telegraphic technology helped transform this state of affairs. The telegraph and the railroad became indispensable to the commerce of the country because they allowed for the more rapid and efficient movement of goods. For example, Field (1992) indicates that there were some industries that benefited extensively from the logistical control brought about from the union of these two industries. The meat, ice, and fruit and vegetable industries (circa 1874) benefited due to the reduction in storage costs made possible by reliable timetables for train dispatches, as well as the savings incurred from the ability to control "shelf-life" of their perishable products. Field (1992: 410) writes that "Logistical control of distribution allowed for greater profitability for owners."

The centrality of these two technologies or industries to commerce can also be viewed from a later vantage point. Roy (1983) writes about the evolution of the non-financial interlocking directorate structure in the United States between 1886-1905. Figure 2-3 is a reproduction of Roy's interlocking directorate structure, 1893-1897. We see that the railroad and telegraph industries (along with coal) formed the core. The railroad and telegraph had 28 shared members of their boards of directors. The railroad and the coal industry had 24, and the telegraph and coal had 3. Roy tells us that these three industries were bound, not only by common ownership, but also by interdependence. Additionally, Roy states that some industries were "preferred customers" and stating that meat packing was one such industry, receiving rebates or discounts.

Figure 2-3. Roy's Interlocking Directorate Structure 1886-1905

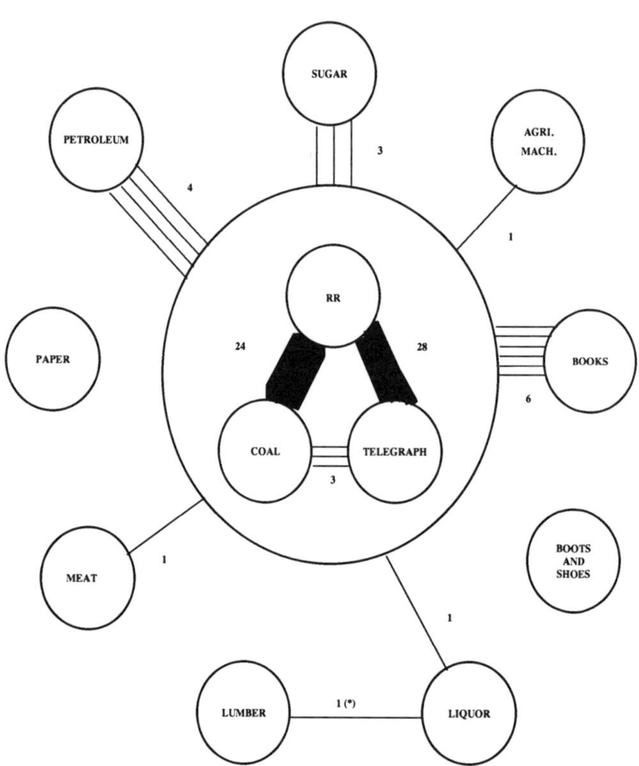

*** Numbers represent number of connecting lines between entities**
Each connecting line = 1 shared member on the Board of Directors

Decline

Telegraphic innovation became embedded in the accumulative cultural knowledge of American society, thus itself becoming part of its own diffusion process at later points in its history. As the idea of instantaneous communication became an integral part of American thought processes, it became an important tool for the organization of various aspects of social life. As telegraph technology continued to diffuse, new needs and applications were found for it, further increasing the diffusion. It is also important to note that incremental improvements were also being made to telegraphy so that the technology being diffused in the 1870s was not the same technology that was diffused in the 1890s, nor was the society in which it was being diffused the same. To say that it was not the exact same technology that was being diffused in later years is to recognize that incremental improvements, such as the number of messages that could be sent over a wire, imply that, on a certain level, a new technology was actually entering into the innovation process, even though we tend to treat a technology as if it were the same throughout its history. Betz (1993) notes that several next-generation technologies can be enveloped by an overall S-curve, suggesting that one must be sensitive to the effects of incremental innovations. For example, an examination of the s-curve (Figure 2-4) for telegraphy actually shows two other downturns before the one caused by the introduction of another radical technology which was to replace the telegraph. These downturns were primarily responses to general economic and strikes against the telegraph company in 1883 and 1895, but also represent periods in which the original single line technology was constantly becoming more complex (that is, due to the introduction of duplex, then quadruplex lines).

Telegraph technology, even as it continued to spread, became an antecedent technology for the evolution of the next major innovation in communications, the telephone, which was responsible for the downturn in the s-curve (Figure 2-4) which we see occurring around 1902. In the course of making incremental improvements to telegraphy (and, I should point out that this process of continuing to improve an existing technology points to the reciprocal and continuous nature of the innovation process) several researchers (Hemholtz, Bell, Gray) came upon the idea of transmitting voice over wires, adding another dimension to the idea of instantaneous communication. It can also be said that the telegraph, rather than solving the need for human agency in the control of communication, spawned the need for better and more diverse means of instantaneous communication. This reflects a general need and consequently, the history of the evolution of telephone technology has a similar trajectory to that of the telegraph through the initial stages of its history. Telephone technology was also a radical, as well as macro-invention. Ownership from the outset was in private hands. The government did not attempt to take over telephone technology until World War I. Like the telegraph, the telephone witnessed a series of competing companies which arose (one of which was Western Union) with incompatible instruments. The final outcome was the emergence of the Bell Telephone Co. as the dominant company in the industry.

The transmission of the human voice was not a part of cultural knowledge. Consequently, it too, only had a pool of amorphous potential adopters. Need for this new technology had to be created, just as it had to be for telegraph technology. A market had to be found, and it wasn't until the 1920s, as Fischer (1992) indicates, that telephone diffusion accelerated quickly. The Bell System finally found the right marketing tool, "sociability" (whereas the telegraph was primarily for business and government uses), hence a pool of adopters was identified. Consequently, when one compares the S-curve for telegraph technology with that of telephone technology, one sees a similar initial, slow ascendancy for each. (See Figures 2-4 and 2-5)[8]. The dotted lines on Figure 2-5 indicate the diffusion of telephone technology as in a series as well. The first decline comes in the 1930s as a response to the depression, and another one (though much smaller) occurs as a response to World War II. Post World War II, amidst the prosperity the United States enjoyed, we see the dramatic increase in telephone diffusion.

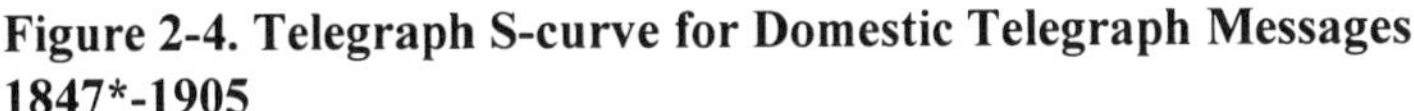

Figure 2-4. Telegraph S-curve for Domestic Telegraph Messages 1847*-1905

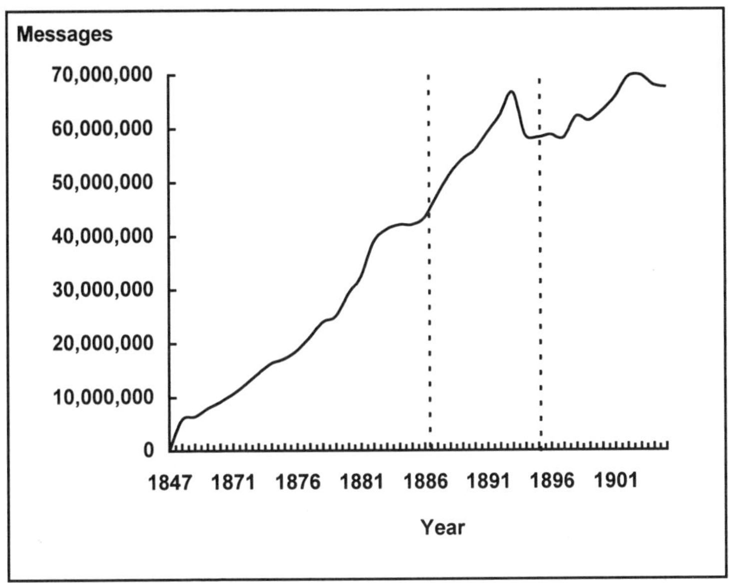

Another form of communication technology, the facsimile, as has a similar initial trajectory, and though its S-curve is not included here, one can infer that it would have a similar appearance in that it was invented in 1842, came into commercial use in 1865, but had a really slow diffusion process until the 1970s. Though the variables that account for the slow initial diffusion of these technologies are particular to each technology, there is a pattern to this diffusion rate that is related to their having similar trajectories through the first three stages (Need, Research, Invention and Development) of the innovation process.

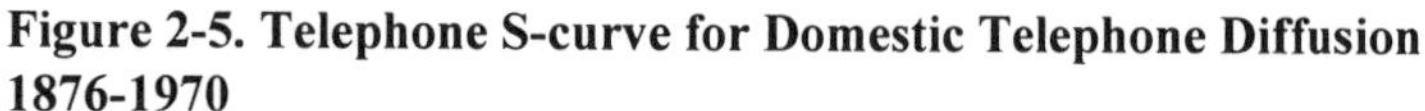

Figure 2-5. Telephone S-curve for Domestic Telephone Diffusion 1876-1970

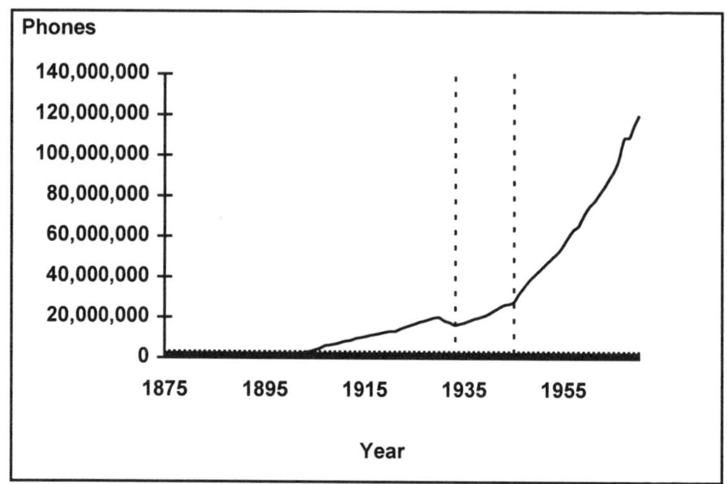

Thus is the trajectory for interpersonal communication technology through the diffusion process. Private ownership of this new, radical invention made it a tool of business. The initial diffusion was slow due to the emergence of many competing firms with incompatible equipment. It wasn't until the systemic needs of telegraph technology were met via the emergence of Western Union as the dominant company in the industry (aided by the railroad industry) that we begin to see the large scale diffusion of telegraphy. At this point in the diffusion process, that is, once the infrastructure had been built, diffusion agencies became strategic in advancing the spread of telegraphy. The government, through the auspices of the military, was responsible for much of the diffusion, especially along the eastern seacoast and along the western frontier. The simultaneous march of the railroad and the telegraph across the United States succeeded in unifying the country in terms of a systemic communication and railroad network.

The S-curves for telegraph and telephone technology illustrate the fact that due to incremental changes in technology, and the impact of social and cultural forces, these interpersonal communication technologies actually diffused in a series of stages. The final downturn in the S-curve occurs when another communication technology replaces the current one. Neither Rogers (1983) nor Robertson's (1981) diffusion models appear to be sensitive to the possibility that a technology can diffuse in a series, that it won't have one historic trajectory but actually several. For example, Rogers explains the decline in the s-curve by pointing to the traditionalists for whom adoption lags far behind awareness and knowledge thus slowing down the innovation diffusion process. Robertson posits the decline as due to sufficient numbers of users giving up a technology or switching to an alternative, and thus the curve begins to fall. By treating the technology as if it was the same over the course of its history (that is, ignoring incremental changes) and by downplaying the historical occurrences (that is, the impact of cultural and social vagaries) some of the richness of detail is lost that would lead to the conception of a series of stages as opposed to a unitary experience of a technology. Betz (1993) suggests that by focusing on an overall s-curve the potential exists for masking the idea of natural limits to phenomenally based technologies. Discontinuities in technical progress for a technology can occur when alternate technical processes can be used in inventions for the technology. Additionally, he cautions us to be aware of how progress altered the morphology of the new technology. And, as Gold (1980) suggests, we need to be able to trace the time path of the major technological innovations as they undergo change.

NOTES

1. This quote was taken from an unofficial sketch of the growth of the Western Union Telegraph Company during the 50 year period 1851-1901, dated June 15, 1901 the sketch was entitled *Western Union Telegraph Company 1851-1901* p. 3

2. See Thompson (1947) for an account of the early history of telegraphy in his book *Wiring a Continent*, as well as Reid (1879) *The Telegraph in America*.

3. The same could be said for telephone technology. Ownership of and control over the physical wires that allowed for the transmission of telephone service was monopolized for similar reasons. Even with the breakup of the Bell System, A.T.&T. still retains exclusive ownership of all the domestic physical lines in order to provide smooth uninterrupted service.

4.The information in this section concerning the military and the telegraph relies heavily on the work of Robert Rupp (1990), and I am indebted to him for making it possible for me to obtain a copy of his book as it was printed in limited edition, as well as his willingness to engage in conversation.

5. As early as 1853, it was recognized that the telegraph could be an important element in national defense. "The improvement of harbors and the dependent rivers, in connection with the existence of railways and telegraphs, not only promotes the interest of commerce, but contributes directly to the defense of the country, by affecting commodious havens for the operations of the navy, and by enabling men and military supplies to be collected promptly and moved rapidly to points threatened by invasion." (National Telegraphic Review & Operators Companion 1853, pp. 24-25). Rupp (1990, p.2) tells us that "Improved communication was always a matter of importance to the army."

6. Axline and Stegengal (1972, p. 12) define the state "as comprising three essential elements: clearly demarcated boundaries describing the territory as a state, a population that more or less permanently inhabits this territory, and a government that exercises ultimate power, or sovereignty, and to which the people accord permanent allegiance."

7. The military was dependent on the telegraph as a means of forewarning the settlers of Indian attacks, as well as enabling the military to better defend against such attacks. For example, Rupp (1990, p. 123-124) cites a military report from Wickenburg, Arizona dated 1879 which states: "The utility of the telegraph line was demonstrated during the military operations against the Chemjarevis Indians in May [1878] last, when this line became the principal medium of communication between the department commander at Prescott and his subordinates and agents in the field and at this point. Full instructions and lengthy detailed reports of operations were exchanged at short intervals, enabling all parties to act with promptness and intelligence, and thus avert what threatened to be a bloody Indian War."

8. The source for these graphs is *The Statistical History of the United States from Colonial Times to 1970.*

Part Two

This section is concerned with the impact of new technology. The result of the successful diffusion of a new technology is that it becomes embedded in the social structure and alters relations within it through this process. An invention, such as the telegraph, which mediated the effects of distance for the first time in human history, had an impact that rippled beyond people's ability to communicate with each other at a distance. Over time, it affected the ways people interacted with each other, the way they thought, and the way they worked. This section examines the impact of telegraphy on several aspects of social life: formal organizations and work, the political system, cultural norms, the legal system, social stratification, and community. While some of the effects of telegraphy were dramatic, others were much less so. And, as testament to the creativity of the human imagination, some of the uses to which telegraph technology was applied, may even be considered novel. However, a variety of each of these types of impacts are presented in order to depict the extent to which new applications and uses for a technology that has become part of the social structure can reach. It is the combination of the dramatic impact and the nuances that subtly alter social relations which leads to social change. Consequently, it is important to highlight both.

III

Telegraph and Formal Organizations

The government's refusal to buy Morse's invention and patents left the development of the telegraph industry in private hands, thus stamping the major direction for this new technology. The telegraph was primarily developed as a tool of business, and only secondarily as a social medium. The changing economic environment of the mid-1800's shifted business concern to the speed of processing and emphasized the need for greater control. The evolution of the first modern business monopoly, and the adoption of the bureaucratic form of organization by the private sphere were responses to the demands of speed and the necessity for control.

The first part of this chapter focuses on Western Union as an example of the first modern business monopoly, and on the Associated Press which, through its alliance with Western Union, came to dominate domestic wire news services, effectively creating a monopoly in the domestic news industry. The economic idea of a natural monopoly will be again explored in relation to communication technology, as it was suggested in the diffusion model in chapter 2 that the telegraph was prototypical in this respect as well. For example, the organization of the Bell Telephone system was modeled after that of the Western Union Telegraph Company.

The second part of this chapter focuses on the telegraph and the rise of the modern business bureaucracy. Though the bureaucratic form of organization was not new, its application to the private sphere of business was. This too, was prototypical for the organization of modern, large scale organizations.

WESTERN UNION AND THE EVOLUTION OF THE MODERN BUSINESS MONOPOLY

As previously stated and elaborated in Chapter 2, it became clear in early 1845 that the government would not take control of his invention, so Morse set about finding the best means of developing telegraphy through private enterprise. The result was the proliferation and then consolidation of the many telegraph companies.

Fierce competition among the telegraph companies resulted in the "Treaty of the Six Nations" in 1857, which effectively carved up the national market across the United States among the six most powerful companies. According to Robert S. Harding (1990: 3) these were: "The American Telegraph Company (covering the Atlantic and some gulf states), The Western Union Telegraph Co. (covering states north of the Ohio River and parts of Iowa, Kansas, Missouri, and Minnesota), the New York Albany and Buffalo Electro-Magnetic Telegraph Company (covering New York State), the Atlantic and Ohio Telegraph Company (covering Pennsylvania), the Illinois and Mississippi Telegraph Company (covering sections of Missouri, Iowa, and Illinois) and the New Orleans and Ohio Telegraph Co. (covering the southern Mississippi Valley and the southwest)." The treaty prescribed mutual aid among the companies as well as respect for the sphere of influence of each company. The member companies were also bound to eliminate or absorb any new competition. However, this treaty began to disintegrate within a short period of time, and by 1860 had withered altogether.

Western Union's emergence as the dominating presence in the telegraph industry came with its extension of its lines to the Pacific coast in 1861. Further consolidations took place among the six systems until there were three major systems: the American Telegraphy Company, the United States Telegraph Company, and The Western Union Telegraph Company. In 1866 the final consolidation took place, Western Union exchanged stock for the stock of the other two companies, essentially making Western Union the first modern monopoly in the United States.

Consolidation eliminated wastefulness and gave the company the ability to universalize tariff rates, with the ultimate goal of reducing telegraphing costs to the consumer and leading to greater consumption. This unification ensured their economic viability and profitability. Western Union's acquisition of its competitors and its alliance with the railroads assured the protection of its routes and increased its ability to control the telegraph industry. In an industry heavily dependent on networking ability, the vastness of Western Union's network allowed it to effectively manipulate its rivals to the point where they became unprofitable and either failed or allowed themselves to be bought out by Western Union (becoming either wholly absorbed or a subsidiary). The year 1867 saw Western Union owning/controlling 85,291 miles of wire, with a net revenue of $2,624,919.19. (Annual Report of the President of Western Union Telegraph Company October 11, 1905). That this profitability was attributable to this new structure is attested to by Western Union President Orton in his Annual Report to the Stockholders, 1869, where he states "Consolidation yielded reduction in the rate for both public and private dispatches allowing for profitability." Additionally, he states "Instead of several repetitions of messages between the great commercial centres of the country, as formerly, transmission is now in most cases direct and instantaneous; and the operation of our system over the vast territory covered by our lines is fast assuming the certainty and uniformity of mechanism. Not only, however, have the public gained in time and greatly increased facilities by these consolidations, but they have received the benefit of large reductions in the rates for both public and private dispatches." (p. 9)

The monopoly set up in telegraphy was subsequently copied by the fledgling telephone industry. For telephony, monopoly came via control of patents. Theodore N. Vail, who became general manager of the National Bell Telephone Company in 1879, and who was knowledgeable about telegraphic technology,[1] sought to steer the company to standardization through the control of and development of any technology related to telephony. He sought to gain this control through the use and acquisition of patents. "From the outset he [Vail] undertook to 'occupy the field,' as he termed it, . . . he explained how his company proceeded to surround itself with everything that would protect the business, that is the knowledge of the business, all the auxiliary apparatus; a thousand and one little patents and inventions with which to control and get possession of." Furthermore, Vail stated " . . . we recognized if we did not control these devices someone else would." Noble (1977: 11-12). This monopoly, created relatively shortly after the inception of the Bell Telephone system, allowed it to dominate this field and control its development well into the 20th century (i.e. until the government forced the breakup of the system.[2]

The Telegraph and the Associated Press

The expense of long distance news reporting coupled with the public's desire for rapid and timely information led newspaper publishers to consider some sort of alliance. A careful review of the situation convinced editors that some sort of a New York press association was needed." These editors believed that such an association, working in alliance with other presses in the North and South, would strengthen the newspaper industry and allow it to overtake the U.S. mail in the delivery of news. Clearly, an association of presses in conjunction with the telegraph seemed to be the answer to the newspaper industry's problem of instantaneous news reporting.

By 1848, two associations were formed in New York: the Harbor News Association, dedicated to obtaining foreign news, and the New York Press, dedicated to the collection and dissemination of domestic news. Subsequently, other regional associations formed (such as the New England Assn., the Southern Assn., Western Assn., and the New York State Assn. (Thompson 1947: 235)), and eventually all these taken collectively came to be known as the Associated Press.

Telegraph companies entered into alliances with newspapers. Thompson (1947: 229) suggests that the individual/company that controls the transmission and distribution of news over a particular line would have a "monopoly at once powerful and profitable." Control leads to the ability to dominate and effect policy decisions. Though the early relationship between the telegraph industry and the newspaper industry was turbulent, by the 1860's, the Associated Press and Western Union had a firm alliance through which they dominated domestic wire news services. This control over information allowed the Associated Press to become a very powerful news agency. The power and control this alliance yielded were alluded to in the 1870 argument by the government for the establishment of the Postal Telegraph Co.[3] The contract between Western Union and the Associated Press was thought to be against public policy because it stipulated that the Associated Press members would not "advocate the establishment of competing lines to the injury of the business of the Western Union Telegraph Company." Orton (1870: 16). Mr. Washburn, a spokesman for the government, implied during these same hearings, that, in return, the Western Union would establish favorable rates for the Associated Press and less favorable rates for non-members. The way the Associated Press was said to control the establishment of new newspapers is through membership in its association. Fees for transmission of news are set at a price, available to anyone. However, where an independent newspaper would have to bear the whole cost, the members of the association each only paid a fraction of the entire cost. Thus, news gathering was less expensive for member of associations and costly for independent papers, or these new papers who did not join an association simultaneous with their inception. Orton denied these implicit accusations saying that the telegraph lines were open to all papers, there was no abuse going on—only business, that is, Western Union was interested in getting as much as it could for the use of its lines, and newspapers were interested in paying as little as possible for that use; arrangements were made. Though this ended the discussion of monopolization by Western Union and the Associated Press at these proceedings, the matter wouldn't die. The Associated Press and Western Union were monopolizing the wires to such a degree' the International Telegraphic Union complained in 1894, that no new newspapers could be established without the consent of these giants because of the their ability to withhold news dispatches.

The histories of the Western Union Telegraph Company, the Bell Telephone system, and the Associated Press illustrate the idea that innovations in communication technology have needs that seem to be best served by monopoly, at least in the initial stages of development. That the telegraph industry developed into a monopoly may have been inevitable due to the needs and nature of its business. Telegraphy can be viewed as a natural monopoly. A natural monopoly is a condition that makes the optimum size of the firm so large in relation to the market that there is room for only one firm. Additionally, the market demand must be such that it could be met by a single firm operating in an area of decreasing costs. Western Union appears to meet these criteria. The preceding discussion points to the optimum size of the telegraph network, and, once the initial poles and wires were in place, increased usage only decreased the costs to the company. Also, the systemic needs of these technological innovations in communication encouraged monopoly. In order for the technology to be effective it had to be able to flow freely to all points connected to it. There needs to be a network. Equally important was the standardization of the "language" being used in this network. Thus, all Western Union offices used the same type of telegraphic equipment in order to ensure the smooth transmission of messages and all equipment spoke the same language, that of Morse code. Prior to this monopoly one company's message had to be reentered manually when it reached the next company's wire.

The Bell system used only Bell telephones, that is, those manufactured by Western Electric, a wholly owned subsidiary of Bell Telephone. The AP, which can be thought of as the first database network, was an information control system that evolved monopolistically in order to efficiently serve the news industry which was critically dependent on timely information. Speaking the same language, whether discussing human beings or mechanical communication technologies, is essential if efficient, exact and meaningful communication is to take place.

The absence of a uniform standard (language) for facsimile transmission was one of the factors that hampered the nascent dissemination of that technology. The history of the fax machine shows a retardation in its initial widespread diffusion. The fax is an example of another technology that evolved as part of a process to improve the capability of another technology. Alexander Bain was interested in developing a system for the transmission of visible symbols over the telegraph wires. He managed to build such a system in 1842. This system needed the refinements and discoveries of future scientists before it would come into commercial use. (See Table 3-1 for a concise history of the early development of the fax machine.) And, a market had to be found for it as facsimile technology was not developed to meet a specific need. The newspaper industry proved to be the market for this technology. There were no exclusive patents for facsimile technology, thus, by 1924 the United States had developed three systems for picture transmission, each mechanically incompatible with the others. Facsimile transmission proved to be so expensive (it could be transmitted only over private telephone lines or radio channels) and slow (about 6 minutes per page) that it did not diffuse widely among businesses or the American public, despite the fact that Western Union had developed a desktop model by 1935,. As a matter of fact, from World War II until the 1980's, specialized applications have been the backbone of the facsimile industry: transmission of news photos, and transmission of weather maps, law enforcement (transmission of "wanted" posters, fingerprints, and police records), and some limited commercial use (expediting customer orders where there were costly changes, expediting monetary transactions) The invention of microcircuits allowed designers to produce small and cheap fax machines which could transmit over ordinary telephone lines. Additionally, the fax machine did not become widespread until the manufacturers of said machines agreed upon a common standard to use concerning the number of scan lines per inch and the number of lines per minute. The inability to (or should I say unwillingness to) standardize transmission language coupled with the lack of technology which would allow fax machines to be made cheaply, resulted in a 40-50 year delay in widespread adoption of this technology. Table 3-2 depicts the later history of fax technology. Now, that is has been standardized and the speed of transmission has dropped to under one minute per page, we are beginning to see widespread diffusion. The facsimile's effects are being felt in many ways, from its traditional business uses, to its use by restaurants and delis, to its invasion of the home.

Table 3-1. Early History of the Development of Facsimile Technology

1842	Alexander Bain (Scotsman) invented a system for the transmission of visible symbols over telegraph lines.
1865	First commercial fax system established in France by an Italian ex-patriot, Giovanni Caselli, who patented an improved version of Bain's device. This system connected Paris with other major French cities and remained in operation for about five years.
1902	Dr. Arthur Korn (German) demonstrated the first practical photoelectric fax system for the transmission of photographs.
1920s	United States began serious experimentation with facsimile technology.
1924	Three separated developed systems (A.T.&T's, RCA's, and Western Union's) were patented and were dedicated chiefly for picture transmission.
1925	A.T.&T.'s telephoto system went commercial.
1926	RCA's radiophoto system went into effect.
1930s	Western Union introduces automatic facsimile telegraph machine which used nonelectrolytic direct-recording paper called teledetos.
1934	The Associated Press took over aspects of A.T.&T.'s picture transmission, renaming it "Wirephoto". Within a year it was so successful that other newspaper interests scrambled to develop picture transmission systems of their own.
1935	Western Union began marketing a desk-top fax machine which took up less than a square foot of desk space.
1937	Introduction of newspapers by radio. The first recorded instance was a newspaper that was delivered by radio fax by station KSTP in St. Paul, Minnesota and then by WOR in New York City. Interest quickly dwindled until after World War II.
1948	FCC officially authorized commercial fax broadcasting. The Miami *Herald*, Chicago *Tribune*, Philadelphia *Inquirer*, and *New York Times* were the first to broadcast special fax editions via their respective FM radio outlets.

Table 3-2. Later History of the Development of Fax Technology 1960-1984

1960	Attempts to reach an international standard began.
1966	First American standard appeared, but it still had compatibility problems.
1968	International Telegraph and Telephone Consultative Committee, CCITT, produced its first attempt at an international standard. Group I faxes could be sent from Europe to America, but not the other way around. Additionally, no one could send faxes in/out of France.
1976	Group II (produced by CCITT) faxes were introduced. They were very expensive and complicated to use.
1980	CCITT produced today's digital standard, Group III.
1984	Group IV, designed to work with the digital telephone networks being introduced all over the world. Speed increased to 6 pages per minute; resolution was also improved.

THE TELEGRAPH & MODERN BUREAUCRACY

The telegraph was not only important due to the advance in technology it represented, but for the organization of work it set a precedent for. Clegg's (1990) work seems particularly applicable to the period of early modernity during which the telegraph emerged. Clegg (1990: 2) tells us that modernity is characterized by increasing degrees of differentiation and division. He states "The crucial hallmark of modernity is taken to be the centrality of an increasing division of labor." Ones experience of work is affected by the culture in which one lives, as well as by the culture within a particular workplace. Changes in the organization of the workplace, as well as, the introduction of new technology, will influence the culture of the workplace, and hence, one's experience of work.

Bureaucracy was not a new concept to society, but the particular form it took and its application to the population at large, and its impact on society was. "In modern America, large-scale organizations are a taken-for-granted aspect of everyday life . . . they dominate most people's working lives. Consequently, Americans tend to take for granted the rules, hierarchy, and paperwork of bureaucracies . . . they seem to remain for many Americans the expected way to organize work." Trice(1993: 51) Blau & Meyer (1987:7) in discussing bureaucracy in modern society tell us that "Modern machines could not be utilized without the complex administrative machinery needed for running industries employing thousands of people. For example, it was not so much the invention of railroad technology as the invention of management that permitted railroads to traverse long distances." This could also be applied to telegraphy—in order for it to become a *system*, it begged a new form of organization. Chandler (1977: 202-3) also tells us that "The speed and volume of messages made possible by the new electric technology forced the building of a carefully defined administrative organization, operated by salaried managers, to coordinate their flow and to maintain and expand transmitting facilities."

Clegg (1990: 10) states that "Organizations are indeed one of the great achievements of modernity." Many theories have been put forth to account for the evolution of the bureaucratic form within modern organizations. Clegg (1990) discusses two of these theories which are particularly relevant to this work.

In discussing the work of Max Weber, Clegg (1990) indicates that it was the needs of a capitalist economy that required the adoption of the bureaucratic form of organization because it was the most efficient. Clegg (1990: 33,36) interprets Weber to say that rational calculation would limit uncertainty and that "Rational discipline would permeate all authoritative relations of modernity." Additionally, this type of organization would impact on the individual. Clegg writes "The outcome of this process of rationalization, Weber suggests, is the production of a new type of person shaped by the dictates of modern bureaucracy. Such a person, whether in business, government or education, is one with a restricted, delimited type of personality. Characteristically, this is the specialist, the technical expert who increasingly, Weber feels, will come to replace the ideal of the cultivated person of past civilizations." Certainly this was true of telegraphers. The organization of Western Union prescribed the "proper" behavior for the telegrapher. Such rationalized telegraphers eventually replaced the pioneer "characters" of early telegraphy.

An alternative view to Weber's is presented in a summary of the work of Williamson. Clegg (1990: 67), in discussing Williamson's work, states that "{B}ureaucratization occurs when transaction costs can be minimalized by internalizing them within firms rather than mediating them through markets. In this perspective, first there were markets and then there were organizations which grew as a response to market failures." Clegg further explains that buying out competitors, suppliers, marketers and so on was a way of avoiding the uncertainties of external markets. And, along this same line of thinking is Chandler's view which Clegg(1990: 69) represents through the argument that "they grow where they are market driven".

Either of these models could be applied to the adoption of the bureaucratic form of organization by the telegraph industry (specifically, Western Union). However, I believe that Clegg's (1990) contingency model, in concert with Granovetter's idea of embeddedness,[4] seems most relevant since it is in keeping with the overarching idea of this work that there is a dynamic relationship among people, technological innovation and culture. This perspective advances the idea that conditional events which determine organization structure are consequences of growth in organization size, constrained by the effects of culture. Clegg (1990: 153) informs us that "Organizations are human fabrications. They are made out of whatever materials come to hand and can be modified or adopted. Organizations are concocted out of whatever recipe-knowledge is locally available." Additionally, he tells us that "In the early stages of modernity it was quite natural that organizations should have been built out of locally available material. Fabricators of organizations would have drawn from the material culture of their immediate institutional environment." This perspective would lead one to look at the institutional models and the cultural values that were available and dominant during the period that telegraph industry was undergoing organization.

Nelson (1982) discusses the political ideology that was dominant during the time the telegraph industry was emerging (1830-1870). He traces the transformation of government from a party-centered to a bureaucratic system of authority. Key to Nelson's work is the concept that the United States was/is a pluralist society, and that this pluralism is reflected in its institutions. Consequently, Nelson instructs us, there was a concern in the political sphere in the mid-to-late 1800's with establishing a decision-making process in accordance with moral, apolitical (that is, not party-centered) and professionally administered standards. The bureaucratic form of organization was viewed as an expression of this pluralistic, egalitarian ideology because it could mediate the effects of discrimination based on social characteristics. When one looks at the emergent structure of the telegraph industry from this perspective, one sees that the economic system of laissez-faire capitalism coupled with the pervasive cultural ideology (as exemplified in Nelson's (1982) work) of the period, may have led the original "fabricators" of the Western Union Telegraph Company to look to local models of organization that appeared to work with large scale enterprises that had systemic needs. The military model of bureaucracy was one that was most readily available and proven.

The military model of bureaucracy informed modern bureaucracy mediated through telegraphic technology. Chandler (1977) tells us that the managers of the first large railroads did not borrow directly from the procedures of the military. He says that the military model may have had an indirect impact on the beginnings of modern business management. However, if one looks to the management of the Western Union Telegraph Company, one can see a much more direct impact of the military model.

Chandler (1977) says that a number of West Point graduates were involved in the early building and management of railroads. This was logical in that the best formal training in engineering in this country up until the 1860's was provided by the United States Military Academy at West Point. The training these men received provided the rational, analytical skills necessary and useful for managing modern business.

Gabler (1988) disagrees with Chandler and sees a more direct relationship. Gabler (1988: 46) was emphatic about the association between the military and the telegraph business: "Because the telegraph, like the railroad, was a form of enterprise so unlike the traditional small-scale ones of workshop or merchant's office, there was but one model, as Harold Livesay notes in connection with the railroads, that could bring rational structure and discipline to the new corporate giants, and that was the military one." This model was not necessarily right for all types of businesses, but does appear to be appropriate for large scale businesses with systemic needs. A listing of some of the similarities between the military model and the organization of the Western Union Telegraph Company would be illustrative.

Western Union referred to the physical breakdown of its various units as "Divisions". For example, in 1870, the organizational breakdown of Western Union included 3 divisions: Central Division, Eastern Division, Southern Division, and the management structure was exactly duplicated in each division. This uniformity of operating procedure nationally, was something new for the working people of nineteenth century America. Furthermore, company directives came down as "general orders" and "special orders". Operators in large offices Gabler (1988) tells us, were grouped into "squads"; *uniformed* messenger boys were called by number and placed under "sergeants". And, according to Gabler (1988), Western Union had a newsletter, one section labeled "The Service", listed such things as monthly transfers, appointments, dismissals and resignations. Western Union also issued a rule book detailing the rules governing how the company's business was to e conducted, as well s rules governing the conduct of employee's behavior. Clearly, the nature of Western Union's business made a more direct application of the military model efficient for managing its business. However, it should be pointed out that Gabler was not suggesting that Western Union's organization was simply a transposed military one. He recognized the dynamic nature of industry and management. Gabler (1988: 46-7) states "What was likely at work was a kind of management dialectic: the army had things to offer those interested in corporate empire building, but telegraphy and railroading, of necessity, themselves spurred management innovation. The two fed off of, and influenced each other."

What is undeniable however, is that telegraphy, via Western Union, brought bureaucracy and its attendant relations of labor and management, into the private sector. Alexander James Field (1992: 413) reinforces this point when he writes:

> "The development of logistical control by telegraph also affected the character and practice of business management in those sectors that could benefit from it. In the decades prior to the second half of

> the 19th century, management consisted almost exclusively of squeezing effort from generally recalcitrant labor. Productivity per unit of effort increased over time because of the implementation of labor-saving technical change, a process in which entrepreneurs played a role. But day-to-day enterprise management, where practical, remained almost exclusively pre-occupied with the extraction of effort. The magnetic telegraph offered a new means of cutting costs and increasing total factor productivity."

Western Union as the Prototype for the Modern Business Firm

Western Union, itself, can also be thought of as the prototype organization for modernity. Where industrialization has been credited with destroying people's intimate identity with their labor by removing it from their control, the institutionalization of the bureaucratic form of management (as exemplified by Western Union) influenced the way people conceptualize work and self-identity (two concepts very strongly related in American culture). Such alteration in the relations or work enable business and the government (in as much as it is organized along the same lines) to increase their control over the individual (who in turn loses power and control). Taylor's (1985: 3) writing seems very apropos: "Perhaps the most persuasive of all rationalization is the increasing tendency of modern society to regulate interpersonal relationships in terms of a formal set of impersonal and objective criteria."

Weber indicates that one of the four basic factors of bureaucratic organization is the existence of a set of rules, so that everyone can adhere to the same standards. Clearly, this was the case with the Western Union Telegraph Company. What we witness, via the Western Union Telegraph Company, is that once this rationalization process began, further bureaucratization and specialization evolved. We can see evidence of this by comparing the book of rules published by Western Union at three points in time, 1870, 1884 and 1900. In 1870 there were 105 rules broken down into 5 subsections: receiving dept. (rules 1-17); operating dept. (rules 18-40); free messages (rules 41-51); delivery of messages (rules 52-60); general rules (61-105). Rules dealing with accounts (64-72) were subsumed under general rules.

In 1884, there were 100 rules broken down into 5 subsections: Receiving dept. (rules 1-25); Operating dept. (rules 26-45); Delivery dept. (rules 46-61); Accounts, Reports and Remittances (rules 62-84); Miscellaneous (rules 85-100). A comparison of these two rule books shows that while the overall number of rules declined slightly, they concurrently became more specific and specialized. For example, free messages (1870), rules 41-51 became, in 1884, one rule, number 79, incorporated into the rules governing accounting. Also, if one were to look at the rules governing the delivery of messages one finds an increase in the number of rules (9 vs. 16) as well as an increase in specificity—for example:

Rules, 1870	Rules, 1884
52. All messages must be promptly delivered to the person addressed, or, in their absence, to their agents, clerks, or some member of their family, and a receipt taken for the same. In all cases follow strictly the directions in the message as to *place* of delivery. If, however, a message which is addressed to a person at his place of business is received after such place of business is closed, it may be delivered to him at his residence; but unless delivered to him personally, a duplicate of the message shall be delivered at his place of business the next day.	53. Messages are not be left with unauthorized persons. A message must not be left with a janitor or porter of building for delivery by him, nor be slipped under a door, nor left in a letter-box, unless the addressee has *filed* with the manager a *written* request for such delivery; nor will a messenger allow any unauthorized person to know to whom a message is addressed. [A separate rule deals with the situation in which a business is closed.]

Comparing these two rules, we also see an increased emphasis on formal record keeping—" . . . *filed* with the manager a *written* report."

The rules are further enhanced by the use of standardized forms. For example, Rule 53 (1870) states "In case of non-delivery, the reason in writing must be given by the Messenger to the Delivery Clerk, who will notify the office from which the message is received, and will also make a record of the same upon the envelope of the undelivered message." In contrast, Rule 55 (1884) concerning notice to addressees of undelivered messages states "When a message cannot be delivered because the addressee's place of business or residence is closed, or because no authorized person can be found to receive the message, the messenger will leave a notice (Form 66) at the place of address."

The combination of rules and standardized forms was an attempt to remove any possibility for people to act as individuals or with creativity, as well as ensuring that messages arrived in tact. Job performance became routinized in the interest of efficiency. Noble (1977: 15) tells us "In short, rationalization might be defined as the destruction or ignoring of information in order to facilitate its processing. . . . One example from within bureaucracy is the development of standardized paper forms." By 1884 the number of standardized forms used by Western Union also increased. Additionally, this standardization and specificity was in part motivated by legal issues and controversies in which the telegraph was becoming embroiled. (For details, see Chapter 7 on the telegraph and the judicial system). The company was liable for its employees actions.

From 1900 onward, we see even further specialization. Western Union ceased to publish a single rule book for everyone, and instead published rules for specific jobs; Rule and Instructions for the Government of Foreman and Lineman (1900); Instructions for Foreman and Division Lineman (1902); The Western Union Telegraph Company Rules for Construction and Repairs (1905); Money Transfer Service Rules (1915); Time Service Manual for the Solicitor (1915); The Western Union Telegraph Company Specifications for Underground Construction (1922).

Specialization took other forms besides formal rule books and standardized forms. The appearance of the auditor in Western Union hierarchy is very important. The increasing size and complexity of Western Union demanded management innovations. Bell's (1973: 29-30) concept of "intellectual technology" can be applied to this phase of telegraphic technological development manifested in Western Union's changing administration needs. Bell writes:

> "An intellectual technology is the substitution of algorithms (problem solving rules) for intuitive judgments. These algorithms may be embodied in an automatic machine, or a computer program or *a set of instructions* [I added underline for emphasis] based on some statistical or mathematical formula; the statistical and logical techniques that are used in dealing with organized complexity are efforts to formalize a set of decision rules."

Reid (1879: 554) tells us "In some important respects the department of audit is the one on which the executive management of the telegraph has largely depended for the care of its administration. Nothing has been so safe a guide as the logic of statistics. (Dependence on statistics was seen as more scientific and more certain.) When, therefore, the business in 1866 began to assume national dimensions, the office of the auditor became one of prime importance." Western Union had a fully organized auditing department by 1867, and this reflected the first addition to its hierarchy since its inception.[5] It represented a form of rationalization that further increased centralized control.

The desire to bring the administration of Western Union up to the needs of this new national corporation required more specific departments (such as auditing) to be organized with carefully delimited responsibilities and authority. What Western Union also needed was an emphasis on the "science" of telegraphy in order to gain more control over the rapidly expanding industry. Reid (1879: 533) tells us "Marshall Lefferts had done much to show the value of statistics, and had laid down important ground-work for systematized and scientific telegraphy." However, it was not until President Orton's tenure that electricity, as in science, was seriously acted upon. By 1873 we see the next major addition to Western Union's organizational structure—the Electrician (who was the equivalent of today's Electrical Engineer). It was from among the ranks of these electrical engineers that further improvements and innovations in telegraphy were to come. In essence, the establishment of this department within Western Union can be seen as the first corporate research and development division, characteristic of modern organizations. The establishment of this corporate division also impacted on the innovation process by adding another dimension.

Western Union Telegraph Company was also one of the earliest conglomerates. One can hypothesize that the motivation for Western Union's diversification into manufacturing was directly related to its needs for the telegraph system as well as for increased profitability. Though Chandler (1977) indicates that this "make or buy" decision doesn't always favor "make"; it did, however, during the initial development of TWUTC. For Western Union, the increased control over having its needs met, meant more total control of the industry. Western Union's President Orton wrote, in his Annual Report to the Stockholders (1873) concerning the factory Western Union built on Church Street in N.Y.C.;

> "[the] building is admirably adapted to the manufacture of telegraph apparatus, and is in close proximity to the permanent headquarters of the Company in their new building. The factory is now capable of supplying all the apparatus required by the Company, and its capacity can be more than doubled when required. The operations of the factory for the past six months show a small profit after deducting the interest on the investment. The apparatus made at this factory greatly excels any other manufactured in this country, and it is the superior quality of the material and workmanship, rather than the saving in the cost of the instruments, which constituted the great inducement for establishing it." (p. 19)

The ability to control the externals that impact on a company means greater overall control, thus increasing the power and position of that company to the degree that it makes them independent. In the long run, such consolidation can lead to greater profitability because they can control quality and supply of necessary materials.

The history of Western Union can be considered prototypical of the modern bureaucratic firm. Western Union's hierarchical organization, sets of rules (which today can only be rivaled by the United Parcel Service), specialization, and the general orientation to the "administration of things", is something quite familiar to most modern Americans and forms the culture of many of the large-scale firms in which they labor. That this form of work organization happened here first (or more accurately, simultaneously first with the railroads) was most likely the outcome of this particular technology innovation as it led to the building of a large scale industry with systemic needs.

The Telegraph, Bureaucracy, and the Newspaper Industry

Telegraphic technology also facilitated organizational change within the newspaper industry. James Gordon Bennett (of the *New York Herald*) is credited with organizing the business of news gathering and editing along the lines which remain standard today. (Tebbel 1969). Bennett became famous for his use of the telegraph to help him outperform his rivals. The speed with which a newspaper was able to gather and disseminate news became crucial to its economic viability and profitability after the adoption of the telegraph. Telegraphy intensified the need for coordination and control within the newspaper industry. What emerged was the institution of the bureaucratic form of organization, and one of the characteristics of bureaucracy had special relevance for the newspaper industry—that of specialization. For example, though the war correspondent existed as a specialist prior to telegraphy, he was joined by other specialists after the widespread introduction of telegraphy. The need for on-the-spot news reporting foretold the modern accent on reporters having a "beat". In the years after 1846 the field of journalism saw the emergence of the foreign correspondent, as well as the "police beat" and the "city beat" as areas of specialization within journalism.

Standardization, another feature of bureaucratic organization, was crucial to the newspaper industry, especially as it related to distribution. Scharlott (1989: 714) tells us that "Newspaper growth apparently surged nationwide, not just in Wisconsin, following the introduction of the telegraph. In 1852 an Ohio newspaper editor observed that: 'In the newspaper business the increase within the past few years past has been truly astounding, and this is shown, not only by the number of new papers established, but by the large circulation gained by most of the leading papers throughout the country.' [B]y the agency of the telegraph, general intelligence is conveyed with the rapidity of lightening, and published simultaneously in all the leading cities of the Union." Clearly, with the public's appetite for news whetted, and increased competition, efficient and rapid distribution of a newspaper became essential. Consequently, standardized rules for its distribution became the norm.[6]

LATER IMPACT OF TELEGRAPHY

The industry that had to be built, which ultimately translated into the organization that was built, was constantly changing in order to keep pace with the incremental changes in telegraph technology. New applications for this communication technology had to be found in order for the organization, originally built as a vehicle to diffuse telegraph technology, to survive even after the telephone replaced the telegraph as the main technology for interpersonal communication. The nature of the organization changed from one which primarily provided a direct means of interpersonal communication using digital technology which left a record, to manager of digital data communication. Hence, the impact of technology needs to be discussed in historical perspective. The impact that a technology has will change as the technology becomes more widely diffused, as new uses are implemented, and as improvements or innovations are made to the initial technology. The interactions of the adopters with the new technology will also influence its later development. Gold (1980) suggests that one should pay attention to the intermediate effects that an innovation can engender within firms and beyond. Effects of an innovation emerge gradually, are influenced by developmental improvements in the innovation itself as well as by associated and independent technological and managerial innovations, all of which interact with various external pressures -price, demand (including customer needs), and supply.

The previous sections illustrated the ways in which telegraphic technology influenced the evolution of the business monopoly and the bureaucratization of the business firm, concentrating essentially on the earliest adoptions. However, though telegraphy was fairly rapidly replaced by telephone technology as a means of direct, person-to-person communication, the industry did not die altogether. Instead, continuing innovations in telecommunications as well as new market needs generated by the existence of telegraphy colored the direction of Western Union and the record communications industry. As voice communication replaced record communication as the dominant form of instant personal communication, Western Union directed its efforts to marketing other services through the years. Western Union's message services included: telegrams, day letters, night letters, personal opinion messages (those which had to be 15 words or less and had to be addressed only to the President, the Vice-President, Congressional members, governors, lieutenant governors, or members of state legislatures), press dispatches, telex, Tel(t)ex, singing telegrams, hotel/motel reservations, candygrams, dollygrams, flowers by wire, wire fax, telemeter, and telegraphic money orders. In addition to these services, Western Union provided private wire services, For example, Western Union leased closed-circuit facsimile systems to government and industry, and management information services. It is this service that functions as the focus for this section on the later impact of telegraphy.

Massive amounts of information can be divided into two classes: perishable—information that is only useful if it is obtained and processed within a fixed period of time; and non-perishable or long-term—that is, information which represents the steady accumulation of facts with a longer span of usefulness, such as demographic statistics or scientific information. Initially, the telegraph offered a way to give human agency or control to perishable information. In its later years, the telegraph industry shifted its focus to the management of non-perishable information. Western Union shifted its focus to the provision of communication services designed to meet real and created needs of government, industry, business and the public. By 1965 Western Union expressed goal was to become *the* national information utility.[7]It was through Western Union's close relationship with another technological innovation which is also based on digital communication, the computer, that it hoped to achieve this position. Western Union pioneered in adapting private wire systems for data processing. Its equipment and circuits, combined with modern business machines facilitated the diffusion of data processing technology.

In the 1960's Western Union opened a computer laboratory in Fair Lawn, New Jersey which was the only laboratory of its kind where computers of different manufacturers could be tested and programmed and customized for individual clients. It was due to this facility that Western Union was able to become the first provider of computerized management information systems which are now widespread throughout formal organizations of various kinds. By 1966 Western Union had established a Management Information Service which was capable of providing three types of service: total system—which provided the customer with computers for both message switching and data processing, tailored to meet the customer's requirements; communication switching—where Western Union provided the customer with one or more solid state processors (computer switching) for switching messages and data between circuits; shared services—Western Union provided access to its computer situated in a central geographic location for subscribers to use. If the customer required, this system had the capability of being expanded to form a network of several computers interconnected to each other for rapid exchange of data or information. Thus, Western Union was able to write in its report to the Office of The Director of Telecommunications Management in Washington, D.C. in 1966: "Having pioneered in the custom-design of leased communications systems to meet the specific requirements of individual business firms, Western Union today can furnish complete tailor-made management information systems to meet any customer's total requirements—including accounting, cost control, production planning and scheduling, inventory control and distribution, planning and control, and other management functions. The Company will plan, install, and service a total management system for any firm, large or small." (p.52)

Western Union's installation of its INFO-MAC (Information Multiple Access Computer Service) in 1965 marked the beginning of its goal to establish a national information utility (a predecessor of the Internet, if you will). Figure 3-1 depicts Western Union's Integrated Record Communication Network. This system integrates all of Western Union's services and systems into a highly flexible record communication/information system which interconnects with non-Western Union systems. It is important to note however that, to date, no one company has been able to dominate the Internet.

Figure 3-1. Western Union's Integrated Record Communication Network

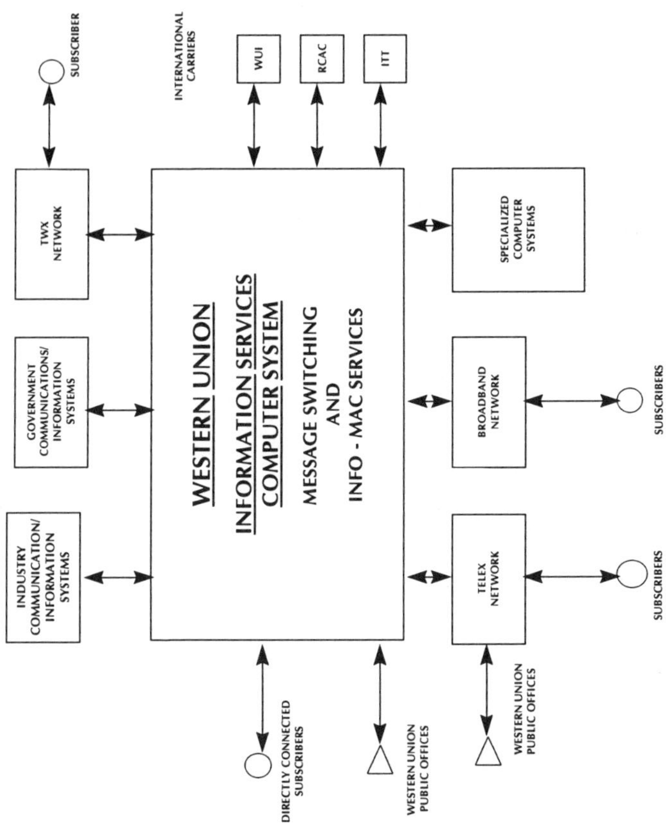

There are a few other technological services that Western Union was instrumental in pioneering (and this should not be surprising since

Western Union was well placed due to the existence of its network) that have become staples of the modern formal organization. First, as indicated in an earlier section of this chapter, Western Union was the first to develop a desk-top fax machine, and due to continued innovations in this technology, fax machines have been widely diffused. Western Union also developed a form of communication that can be said to be the predecessor of e-mail, telex. Telex was an automatic direct-dial teleprinter exchange for instant world-wide two-way telegraphic communication between subscribers in the continental United States, as well as Canada, Mexico, and 112 countries overseas. It was a multiple address service that also had the capability to permit a telex subscriber to send a single common message to as many as 100 telex or TWX subscribers, or any combination of both.

Another of Western Union's services can be said to have been a predecessor of personal computer services available today, that is, home based shopping. The introduction of digital computers into the message switching system allowed Western Union to expand its public message service to include the ability of their computer centers to act upon the information content of messages and perform some of the functions that users normally expected to perform. For example, Western Union's "Operator 25" Service provided access to public data bases which would provide the user with the ability to obtain price information as well as the location of the nearest distributor. Modern computer technology offers the shopper the additional choice to process their order right from their own computer.

The impact of the telegraph on formal organizations varied over time. Some of the impacts were direct and others indirect and evolutionary.

The evolution of a monopolies as exemplified by the history of the telegraph and the history of the telephone suggests that a natural monopoly is likely to emerge in communication technologies due to the systemic needs, that is, the need for free flowing information to all points in the network, of such technologies. The story of the Associated Press also illustrates the way in which other business monopolies can emerge when they have systemic needs, are dependent upon, and working in concert with, monopolistic communication technologies.

Additionally, it is important to note the vital role standardization plays for networked technologies.

The telegraph also brought bureaucracy into the private sphere. Bureaucracy was not a new form of management. It already existed in the public sphere, that is, as a form of administration used by the government. What was new however, was the application, an application brought about by the convergence and interplay of technological innovation, economic need and social organization. The telegraph (and the railroads) brought this form of administration to the modern firm. For example, use of the telegraph brought bureaucratization and standardization into the newspaper industry almost immediately after the newspaper started relying on the telegraph for newsgathering. The bureaucratic form of firm management that was pioneered by the telegraph and railroad industries became the prototype that would dominate management of large scale enterprises well into the twentieth century.

A word should also be said about the ability of bureaucracies to live on after their initial mission is accomplished. Western Union survived as an organization by shifting its sense of mission over time from a vehicle of diffusion of a radically new technology, to finding new, unique applications of the incremental changes in the original technology, to digital data management as technology moved into the computer age.

All the above mentioned changes exemplify clearly, the ways in which the invention of telegraphy had effects that rippled beyond its initial point of diffusion and affected change within formal organizations and the organization of work.

NOTES

1. Theodore N. Vail was the nephew of Alfred Vail, a pioneer and early entrepreneur of telegraphy. Paine (1921, p. 107) says of T. N. Vail's entry into telephone technology "Telegraphy, the Vail tradition, was to be a thousand times outdone." And, furthermore Paine (1921, p. 107) states that "His [Vail] former interest in patents that would revolutionize the world dwindled into insignificance as he contemplated the significance of this one [telephone]." In addition to having a family history in telegraphy, Vail had experience as a

railroad telegrapher and with the post office system. Vail, according to Paine (1921), was credited with rationalizing and systematizing the distribution and sorting system of mails along the railroad lines. Clearly his vast experience with technologies that had systemic needs, aided him in his managerial position within the telephone industry.

2. Though the Bell System was broken up into independent operating companies in *1984*, and rival companies have come about (most notably Sprint and MCI) with attendant allegations of substandard service from these companies, as well as inconvenience when making calls if they have to be routed through competing companies, we have also recently seen the re-alliance of some of the Bell companies, for example, New York Telephone merged with New England Bell to form NYNEX; N.J. Bell merged with Chesapeake & Potomac & Diamond State to form Bell Atlantic; all told, 23 Bell Telephone Companies merged to form 6, and 2 of the originals, Cincinnati & Southern New England Bell, which were not wholly owned subsidiaries, stayed independent.

3. The government wanted to take control of the telegraph industry and put it under the auspices of the Post Office using The Postal Telegraph Company as the vehicle and making it the industry standard.

4. Clegg(1990: 7) tells us that "embeddedness" refers to those relations of "relative autonomy" and "relative dependence" which exists between forms of economic and social organizations and the respective national frameworks of cultural and institutional value within which they are constituted. He says these configurations are "achieved" phenomena. They are socially constructed, emergent, produced and reproduced. Clegg tells us that Granovetter suggests that "one must develop an 'embedded' view of the social organization of economic action." See pp. 6-9 for a full discussion of Granovetter.

5. This new department employed 52 people, 25 of whom were women. Admittedly, they did the most tedious work in the new department, but it was definitely "white collar" work and this was an important factor in attracting women to these positions. (Women's employment in telegraphy will be discussed in detail in chapter 6.) The actual auditing and booking was done by men, the women did what was known as "heck reports". that is, they checked to make sure that the "received" of all offices compared with the combined "sent" and that the charge or check of all offices corresponded with the receipts. (Reid 1879)

6. It is important to note that it was through the widespread dissemination of information and timely news, that the telegraph had one of its biggest impacts on changing how people lived.

7. The primary source for the information in this section is an unpublished document from the Western Union Archive II Collection at the American Museum of History at the Smithsonian Institution entitled "Western Union's Place in Management and Technology Trends in Telecommunications", March 25, 1966. This report was prepared for The Office of Telecommunications Management in Washington, D.C.

8. Carl Townsend (1984:10) supports this point as he wrote "The beginning of electronic mail can be traced to the first electronic message sent by Samuel F.B. Morse in 1844."

IV

Telegraph and the Political System: Power and Politics

The political arena is crucial for understanding the role of technology in society. When examining the impact of a new technology one must pay attention to the relationship between technology and the government with all its various agencies.

Once the capabilities of telegraphy were known and the possibilities exposed, the further development of this technology became the crucial question and arena for struggle. One of the most fundamental decisions that can shape a new technology concern who owns it and is responsible for its further development. The development issue is very culture-specific. The history of the evolution of the telegraph in Europe and America will illustrate this point. Following the typology cited earlier in this work, the telegraph in Europe evolved from the identification of a specific problem—the military wanted improved control over its communication ability in order to gain a strategic advantage during war. Consequently, once electrical telegraphy was invented it had a guaranteed initial adopter, the government. In contrast, the telegraph in America evolved out of a general need for the improvement of communication at a distance. Consequently, the telegraph in America had a pool of potential adopters. The government could have chosen to underwrite and gain control of this new technology[1] and thus have gained the power that came with it, or it could have been left to private industry to take control and direct this new technology. While business was lauding itself for its decision to adopt and take control of telegraph technology, the government was having second thoughts about its decision to relinquish control. This chapter examines the struggle between two primary institutions (the government and the economy, or business) over control of telegraph technology and which would play a primary role in the continued evolution and diffusion of telegraph technology. The resolution of this struggle was that the further development of telegraphy in America was once again left primarily in private hands. A brief discussion of the European experience, where telegraphy was controlled by the government, is relevant, as the comparison becomes part of the later American debates on government ownership of telegraphy. American politicians pointed to the French government's monopolization of telegraphy as a model to be emulated due to the control over national security that such an arrangement afforded. They pointed to British monopolization as the way to ensure democratic access to telegraphy.

The final section of this chapter discusses the role of the government in computer technology as it too, eventually emerged as a communication medium, though not of the same sort as the telegraph and telephone.

FRENCH AND ENGLISH TELEGRAPH EXPERIENCE

In France, once electrical telegraphy had been invented, the government was quick to use it to replace optical telegraphy, which was limited in its ability to effect control. Political pre-occupation with war and the attendant need for better control over information processing meant that the government would be the initial adopter of telegraph technology, and that it would be the main diffusion agency. In the case of France, this meant specifically controlling access to information. Hence, the government acted as the gatekeeper for this technology and the diffusion process.

Kieve (1973) informs us that in most European countries the telegraph system was run as a state monopoly, almost from the beginning. The telegraph was considered to be of critical military and political importance, and no important lines were constructed by private enterprise. For example, Kieve (1973: 46) writes "In France, the first electric telegraph line was constructed by the government for its own use in 1845, and not until November, 1850, was public use allowed, with priority for official dispatches. French culture was attuned to the importance of control of information from a military perspective.[2] In July, 1847, the Minister of the Interior, Locove-Laplagne, declared in the Chamber of Deputies: 'La télégraphie doit être un instrument politique, et non un instrument commerciale.' "

English experience with the telegraph industry tells a slightly different story. There was no guaranteed adopter for this new technology. The government was content to allow private industry to foot the bill for the initial development of the telegraph system. There was no government financial assistance, no central planning of a coordinated national system. In 1840 the Select Committee on Railway Communication commented. "Circumstances may arise in which it may be very inconvenient to leave in the hands of a private company or individual, the exclusive means of intelligence which the telegraph affords, it cannot fail to be of a paramount importance that the government should be furnished with similar means of procuring and transmitting intelligence." Kieve (1973: 36) Thirty years later, in 1870, the English government took over the telegraph system making it a department of the Post Office. Kieve indicates that part of the justification for the government taking over the telegraph industry was based in ideas of morality and fairness. He tells us there was a shift in ideology as the British people became conscious of the moral aspects of many industrial and social problems, and hence we see the influence of cultural values on the innovation process. This new public consciousness was then used by the government and applied to the telegraph industry. The principal of natural liberty, which sounds as it might have been as a system for producing wealth, stood condemned as causing abuses and deficiencies when in came to telegraphic service. The implication was that it did not bring about a just distribution of wealth or profit to the community. Although, the country was governed by the upper class, the commercial classes, Kieve tells us, were prepared to use the state to forward their own interests. Shifting control of the telegraph industry to the state shifted the locus of control to the government, giving it more power and taking it away from the individual owners (business).

These two issues: the "intelligence" importance of the telegraph for military purposes and subsequently, the importance of the control of information flow in general, and the ideological argument of equal access for all citizens, will reappear in American telegraphic history but with different outcomes.

GOVERNMENT VERSUS PRIVATE OWNERSHIP: THE AMERICAN EXPERIENCE

The United States government basically left the development of telegraphic technology in the hands of private enterprise for about twenty five years, by which time it had proved itself to be a commercial success as well as a "necessity" of everyday life. The telegraph had proved itself militarily and politically as well. Consequently, twenty five years after the government said that the best thing would be for telegraphy to be developed privately, it was introducing bills in Congress proposing government ownership and control of the telegraph industry. One has to wonder what prompted this change of policy regarding this technology. Motivation aside for the moment, what developed was a struggle for control and power between two of the primary institutions in society, the government and the economy. And, it is extremely interesting to note that all the voices in this debate had a common rhetoric—its position was based on a concern with the "public" good. It is informative to see how each of these voices, that is, the government and the economy (or business, as represented by Western Union and the newspaper industry) tried to unite its own interests with that of the public's in the hope of gaining support. This argument is one that went on for several years—not until 1874 was the matter finally put to rest. The outcome, however had implications for other communication technologies that followed the telegraph.

The Government's Position

The first bill proposing government ownership of the telegraph industry was presented before Congress in 1866 by which time Western Union had succeeded in creating a virtual monopoly in the telegraph business, having absorbed most of its competition and extended its lines from coast to coast. Western Union had truly become a national company, and the power that this control over information gave them did not go unnoticed, especially as it related to the press. The ability to disseminate all manner of information across the entire United States was entirely in the hands of private enterprise, the telegraph industry and the press. Together they controlled the flow of information, and, except for matters that could be labeled "national security", the government was left out of the control system of this communication network.

Lastly, I suggest that there were monetary motives as well. When the government initially declined to develop the telegraph industry, the industry was operating at a deficit. However, by 1869, Western Union Telegraph Company was generating a profit of $2,748,801.45 (Annual Report to Stockholder, 1905: 6) while the Post Office was operating at a deficit of over six million dollars, according to a New York Article of January 21, 1869. Clearly, the profitability of this industry, combined with the economies that could be brought about by the union with the Post Office, could only, the government felt, put the Post Office Department in a better economic situation.

Consequently, one could hypothesize that the government's growing appreciation of the power and prestige afforded to those who control information (both its processing and flow), combined with the knowledge that the telegraph industry could be run profitably and that the telegraph was of vital importance in military and political affairs all these conspired to entice the Federal government into seeking to gain control over the telegraph industry. The government's main problem was how to convince the American public that this was the right thing to do.[3]

The government's chief arguments can be summarized as follows:[4]

It was in the public interest to allow government control. There were several reasons given for this sentiment.

The Post Office and the telegraph had the same goal—the transmission of information. Consequently, a unity of the two services would create uniformity and allow for cheaper rates. These cheaper rates, would in turn, make telegraphing more accessible to more Americans. In other words, it would make the telegraph accessible to all citizens, not only the wealthy, by reducing costs (reminiscent of a British argument). The government supported their assertions by pointing to the European experience where telegraphy was government controlled.

A sampling of their position statements will illustrate this point.

> " . . . that the post-office and the telegraph have but one and the same object, and that it is for the interest of the people that their management should be in one and the same hands." (House Document #73,: 39)

> "The public instinctively feel that it would be desirable to have the same simplicity and uniformity in the rates for the transmission of messages by telegraph as by mail, with the lowest rates which the nature of the service admits." (House Document #35,: 1)

> "Where the rates are high and facilities poor, as in this country, the number of persons who use the telegraph freely, is limited. Where the rates are low and the facilities are great, as in Europe, the telegraph is extensively used by all classes for all kinds of business." (House Document #35,: 19)

> " . . . the telegraph in the United States is a convenience which our people would be exceedingly glad to use, their tastes and inclinations being in that direction, but which, on account of exorbitant rates, and the general bad management of the service, they are debarred from using; so that, compared to European countries the use of the telegraph in the United States is merely a bagatelle." (House Document #35)

b) Government ownership would break up the monopoly of the telegraph industry, and free the public from the tyranny of such a monopoly. " . . . the press demands a reduction of tariff for its news reports, and a relief from the combined monopolies of the telegraph and the Associated Press." (House Committee on Appropriations, April, 1871) The public was seen as the direct beneficiaries of decreased oppression on the newspapers.

2] The government represented itself as better able to further the development of telegraphic technology because they were not motivated by profit, but public welfare. For example, the government proposed to establish lines where it was deemed necessary, not just where it would be profitable. Bill #1083 states "That the Postmaster General shall establish a telegraphic station at as many post offices along said [telegraph] line as, in his judgment, the public interest . . . shall require . . . "(p. 4) And, House Document #73 states that the interests of the people "demand the impartial extension of facilities."

3] The interests of the Government demand control of the entire wire service so as to be able to conduct public business unimpeded. This also included the ability to properly transmit weather reports, which were very important to American agriculture and commerce, and hence the economy of the country. (House Document #73).

The Press Position

The press came out strongly against government ownership of the telegraph industry. Newspapers felt that government couldn't give them the same level and quality of service. The press maintained that publishers and readers of daily newspapers had a similar interest in seeing that private ownership continued because, under such an arrangement, business was "performed zealously", was marked by "patience and courtesy", and that "The interests of both were identical and reciprocal". The press and the telegraph company each had a business to run, and to do so profitably meant that each had to provide reliable, accurate and affordable service (the telegraph company to the press, and the press, to the public). Each depended on the other to achieve these goals. The press was leery of what government management by the Post Office would mean, considering that the Post Office, and government employees in general, were viewed with suspicion and sometimes contempt, due to the political nature of that employment. A New York Times article of January 21, 1869 asserts "A change of this service from the Company [Western Union], or any other organization like it, to the Government, must be a change for the worse . . . The clerks of the telegraphic branch of the Post Office would surely partake of the spirit which characterizes nearly all American government employees. They would be indifferent, unsympathetic, generally supercilious, disposed to shirk work, and would do not more than was necessary to retain themselves in place . . . Their responsibility would be to their departmental chiefs." Furthermore, this article states " . . . all impelling motive to satisfy the public . . . would be wanting to the bureaucratic managers of the Government."

It seems that such concern about the corruption of the government bureaucracy in the Post Office was not just press hysteria. Nelson (1982) tells us that the practice of awarding contracts for mail delivery primarily to contractors whose politics were acceptable to the administration began as early as 1835 with the tenure of Amos Kendall. Nelson (1982: 25) also states that "Together with post office employees, these contractors and their employees constituted a sizable army whose economic interests ensured their loyalty to federal law and administration policy."

By the time the government was lobbying for the unification of the telegraph industry with the Post Office Department in 1866, the Post Office was the largest branch of the federal government. Nelson (1982: 24-25) tells us that "Its size alone made it important, for by wisely distributing postal jobs an administration could gain power over a large number of employees and their families." Thus, the significance of Post Office Department lay in the control over its direct employees and their families, as well as its control over the private contractors and their dependents, who by 1835 numbered over 20,000, and received over two million dollars from Post Office revenue.

Nelson (1983) tells us that the Post Office was a powerful political machine that gained added importance from the fact that its employees were "agents for disseminating information throughout the country", thereby acting as an agent of social control. The Post Office could help maintain social order and manipulate political opinion, as it did when it refused to deliver abolitionist tracts in the South, because of its control over the flow of information. Thus, each postmaster was a "gatekeeper" of sorts, and, as Nelson (1982: 26) describes them: "Each postmaster became an electioneering outpost."

Consequently, the press believed that the health of their own industry was rooted in private control of the telegraph industry, as it was the surest way to guarantee freedom of information, which was also important to the American public of the 1800's. Additionally, they traded on the popular post-Civil War sentiment of Americans to control and limit the power of the federal government by curtailing the opportunities for political partisanship, reflecting a concern with protecting the rights of all individuals and minorities—the public Nelson (1982). A New York Tribune article (referenced in the *Journal of the Telegraph*, February 1, 1869) states that government ownership is an affront to the "self-helpfulness" and freedom of the American people, and, that "If it were adopted, where would a limit be put on this extension of the functions of Government into to the domain of private business?" Will the railroads be next, was a commonly raised question.

Western Union's Position

Obviously, Western Union was vehemently opposed to the government take over of the telegraph industry as it was a direct threat to its business. The public, again, became central to Western Union's opposition argument. In that unification was an issue before Congress—a group that was supposed to be responsive to public sentiment. Western Union President Orton set out to gain public support for private ownership through personal appearances at public hearings, by speaking before Congress and through the Annual Reports issued to stockholders. Orton designed to garner public support by presenting counter arguments to all the government's allegations.

First, Orton (1870) asserts that the only parallel between the work of the Post Office and the work of the telegraph industry is that they are both means by which ideas are communicated. This, however, is where the similarity ends. One cannot compare the day to day operations. For example, where the Post Office can put a thousand letters in a mail bag, and ship it off to its destination, each telegram is shipped individually. Orton vividly tried to make the point that the operating procedures of each, i.e. mail delivery and telegraphy are totally disparate, to a degree such that few economies could be realized by the union. Postal clerks and telegraph operators were not interchangeable employees. Furthermore, Orton, felt that a comparison of the United States telegraph system with Europe's was unfair. Orton stated that "The telegraphs in continental Europe are owned and operated by the Government, not for the purpose of revenue, but as an element of power; and in no country at the present time is the telegraph under Government control self-sustaining." He stated that just because the European rates were less, it didn't mean that American telegraphy was less efficient in terms of management. American companies had to give their owners a "fair return" for the capital invested in them, as well as pay taxes upon the property and gross receipts of their business. Additionally, Orton (1874) expressed the view that private employees are more reliable because, given that the U.S. telegraph companies are run for profit, their jobs depend on securing results, which means meeting and satisfying the needs of the public. Orton said that government employees have no such motivation to satisfy the public because their pay is dependent more on political partisanship, influence, or favoritism. "The salaries of the officers and employees of the Post-Office are paid as promptly and at the same rates when there is an annual deficit of six million from the operations of that Department, as if the balance were on the profit side of that account." (p. 55) Furthermore, the idea that government could create uniform rates within the U.S. and charge higher rates for transmissions outside the country, is not a fair comparison because the greatest distances messages may travel varies considerably less in Europe as opposed to the United States. For example, "The greatest distance that telegrams can be sent in Belgium is 175 miles, while in America they can be transmitted over 5,000 miles. Additionally, the European countries to which telegraph service was being compared (specifically Belgium, France, and Switzerland) were among the most densely populated in Europe, averaging 250 persons to the square mile, while in the United States the average was 10. "Would it not be absurd, therefore, to demand that our rate should be the same as those of Belgium?" Orton, (1869: 45)

Secondly, Orton objected to the charge of bad management. He put the following question to the Chairman of the hearings in 1870: "It seems proper that at this point I should inquire . . . whether any portion of the people of the United States have, during this session of Congress, or any preceding session within your knowledge, either by personal appearance before committees, by letter addressed to members, or by petition, requested any intervention on the part of the Government in this business . . . " [No reply was made to this inquiry]. Orton suggested that the only time the government should interfere with private enterprise is if it guilty of flagrant abuses. He stated "The only question to be considered is, whether those who control its affairs administer them properly and in the interest, first, of the owners of the property, and second, of the public." Orton (1870: 13) Orton pointed out to Congressional representatives that it had never been necessary in the history of the United States, for the government to intervene "in any of the enterprises undertaken by the people, and in the success of which they are directly interested" (Government Hearings, 1870). Since there were no complaints by the public, Orton inquired as to why the government felt that its takeover of the telegraph industry would be for the public good.

Orton also addressed the issue of press tariffs, stating that, in the past, rates were pricey for the volume of business done, but arrangements had now been made that were equitable for both Western Union and the press. He was very emphatic about stressing the point that the savings incurred by the press were not being passed off onto private customers. He said that reduction in press rates was primarily due to an arrangement where most transmissions were done at night when the lines were not ordinarily busy. Anyone wishing to use the wires at night would benefit from reduced rates. Orton said that the goal of both industries (telegraph and the press) was the dissemination of information, and that this was not only good for business, but advantageous to the freedom of the American people. Orton (1870: 16) stated before the Congress "Ours is a popular Government—our people have an interest in the dissemination of news, while most European governments have an interest in suppressing the news" and again, before the Senate in 1874 (p. 54) that "A Government monopoly of the telegraph might be made far more exacting and oppressive than is possible in respect to the mails."

Clearly, the debate over government versus private ownership was, in reality, a struggle for power and control between government and big business (telegraph and press). And, though much of the debate revolved around the "public", said public was merely a pawn in the battle. The public's ability to control the flow of information had already been coopted, first by the postal system, and later by private telegraphy. The public had already lost power to these two agencies. However, as society underwent modernization, there was a growing realization that public opinion had to play a role in the direction of future development. (Not so much that the public could create or design that development as an original work, as that they had the power to choose among alternative options.) The relationship between the government and the masses (public) had to change. A reading public (made possible by the advances in printing as well as telegraphy) meant that the public was better informed, and better informed meant that they could not be ignored. Consequently, both the government and private industry took great pains to win the approval of the public in order to further their own agendas. Business emerged victorious in this debate, as every bill that was presented before Congress was defeated in Congress. However, one must be careful not to assume that this was an expression of "public" will, that is, of the public influencing their Congressmen. Rather, as Lindley (1975: 128) points out, Western Union used the lobbying of Congressional members to "buttress their freedom from national restraints." The net result was that the power of the economic sphere was increased at the expense of the political sphere under the rubric of two of the values of American culture, freedom and a firm belief in the capitalist system. However, the issue of government involvement in interpersonal communication technology was only to lie dormant, not to be resolved at this point in time.

A nation's communication infrastructure is central to its ability for control, and will influence the very structure of society. Access to and control of communication dictates, in part, where power lies within a society. One gets a sense of the importance of control of a country's information network from the constant reoccurrence of government involvement in communication technology, that is, its attempt to become a vital part of the information processing system. Additional examples will illustrate this point. During World War I the government again took control of the nation's communication systems as a matter of national security. Only this time the government took control of both the telegraph and telephone industries in order to maximize the possibilities of both as complementary and supplementary forms of communication during the war effort. By 1918 the discussion of government ownership was again on the nation's agenda. The point of view expressed by Mr. Vail in his annual report to the stockholders of A.T.&T in 1918 reflects the general sentiment of business towards this current government initiative:

> "The recommendation of the advocates of government ownership to take over the telephone toll and long -distance lines, equip them for telegraph purposes and enter into a destructive competition with the existing telegraph companies for the purpose of destroying their market value and enabling the government to purchase at a low price, is so utterly at variance with any possible standard of public or private commercial honor that it would seem as if the very suggestion would be repudiated."[5]

The "public" was again brought into the argument, just as they had been in the past, with the idea that it was in the public's best interest not to have government ownership of the telephone industry, much less to have government controlling both record and voice communication. Typical of the sentiment as represented by business was the statement:

> "There is little doubt remaining in the minds of the public, but that regulated monopoly is better than unregulated Government ownership, and there is no longer any extensive conviction that there can be effective competition in the electric transmission of intelligence requiring a nation-wide universal system, whether messages or conversations."[6]

Sentiment regarding the telegraph was very similar. According to a *Literary Digest* article of April 19, 1919, during the government's tenure as owner, costs increased, complaints about service increased, and the general sentiment was that " . . . the tendency of complete government operation is toward the worst service for the most money." (p. 15) Again, the public spoke, stating that the government should not get involved in private industries that were servicing the public well because the only effect of government ownership would be to increase the tax burden on the public. *Literary Digest*, (May 10, 1919). Consequently, the government returned the telegraph and telephone to private control after the war.

This debate over government ownership and control was still not over. The next great war brought it to public attention again. Noble (1977) indicates that one of the first things the government did at the outbreak of World War II was to take over the communications industry (among many others) in order to afford itself the utmost control. The ironic thing here is that in this instance, when the war was over, government involvement with the telephone industry actually resulted in an increase in rates and the return of a healthier industry. However, the interests of free enterprise still triumphed. Using the public good as their as part of their arguments, private industry regained control of the telephone industry.

The next big attempt of the government to get control of the latest communication technology came with facsimile technology.[7] During the 1960's, the government attempted to get involved with facsimile service. Interestingly enough, it was the Post Office that thought it could benefit from this new technology. The Post Office spent between three and four million dollars on the project to develop "speed mail"—a system and equipment. Not surprisingly, public sentiment came out strongly against it. One newspaper headline stated "Free Enterprise Attacked Again" *Tulsa World*, (December 3, 1960.) The article criticized the Post Office Department for wasting tax money on experiments with facsimile when "Western Union, a pioneer for 25 years in the facsimile field, had already done all the work." The article ends with a statement and question reminiscent of the debates of the 1800's: "The Post Office Department should be a customer instead of a competitor of existing private business. It now contracts with railroads and airlines to transport the mail. Will it be a matter of time before it decides to operate its own railroads and airlines?" (p. 4)

There was clearly a sentiment against the government increasing its power and control at the expense of the free enterprise system. The *Southern California Industrial News* (January 2, 1961) felt that, the taxpayers would welcome any innovation to cut postal costs or improve efficiency, but it should look to companies already engaged in this service, rather than look to compete with them. "Any other move, it would seem, is a retreat from the concept of government doing for the people only what the people cannot do for themselves." And finally, a commentary in the *Boston Free Press* (December 1, 1960) condemns the government for trying to destroy the private telegraph industry at the expense of thousands of employees and pensioners. It concludes with the statement that "With the present trend toward greater government control, the adoption of this speed mail system by the Post Office Department, will mean one more step toward Socialism." Again we see that government control of an aspect of the national communication system is viewed as antithetical to the public good and a threat to freedom.

THE GOVERNMENT'S ROLE IN THE DIFFUSION OF COMPUTER TECHNOLOGY

The role of the government can be crucial to the diffusion of a new communication technology, especially when it is a new technology that requires infrastructure building that is different from anything that currently exists (that is, that there are no prior models for). Previous sections of this work have illustrated the role the government played, via the military, in the diffusion of telegraphic technology. Fischer (1992: 58) tells us that the government played only a small role in the diffusion of telephone technology.

> "The government's role in U.S. telephony was small. In some ways government modestly encouraged telephone diffusion by holding rates down, pressuring companies to service outlying areas, and requiring interconnection. In other ways government

discouraged telephone diffusion by sustaining Bell's technical standards. But there was no direct subsidy to telephone service; one way or another, users paid the full freight of service and then some."

However, the government's role in the development and diffusion of computer technology is much more dramatic. As with previous communication technologies, the government's role in the development of the computer was through the military. The computer was developed in response to a specific need—the analysis of foreign cryptic messages. Post World War I the Navy directed research efforts toward the development of radar, radio communication and the decoding of cryptic messages. Flamm(1988) tells that in the early 1930's the Navy began to mechanize its cryptanalytical activities with the aid of IBM. It was these cryptological applications in support of national defense that first justified large government expenditures on computer development. Hence, the initial stage of computer innovation had a guaranteed adopter and the technology diffused rapidly through this limited audience. With a few exceptions, (such as the artificial intelligence *software* developed through the Department of defense and the Air Force's SAGE *software* system which was developed to manage large data bases in the 1950s) the government was primarily responsible for the significant developments in computer *hardware* innovation as opposed to software innovation.

Post World War II, as the country switched to a peace time mentality, and the companies such as IBM and NCR were able to redirect their primary attention to the marketplace instead of the wartime needs of the government. Flamm (1988:47) tells us that "By the end of the war [WWII] it was clear that the newly developed electronic computers were going to have a practical impact on business. The idea of applying electronic computers to commercial and business problems became the driving force behind the spectacular rise of the computer industry." Thus, we see that federally supported innovation in computer hardware created a significant technology base for the later development and application of the computer to commercial needs. It was from among the research carried out by the private firms that most of the innovations in software came to meet the needs of the commercial markets. Consequently, the second phase of technology diffusion was addressed to wider, but still limited audience, for diffusion. Cortada (1993) tells us that the public became aware of the computer primarily through the workplace. I think it is important to specify who the "public" is when talking about any technology, for at times the public refers to the mass of people in a society, as in when something is said to be for the public good; at other times the public actually only refers to a small subset of the population. Which public is actually addressed is important when one is concerned about the widespread diffusion of a technology. The public, in the more limited sense of those who came in contact with computer technology through the workplace, were in a sense forced adopters of this new technology; or, in other words, these people constituted a pool of guaranteed adopters. However, in order to view public acceptance in terms the broader definition of public, one would have to measure diffusion by the appearance of the computer with the home unrelated to business use. However, this brings us to a point made earlier, that we cannot discuss the impact of the diffusion of a technology as if it were a uniform and unitary event. One needs to blend economic factors with social factors and take the intermediate effects into account when attempting to understand the impact of technology innovation. Gold (1977: 223-224) indicates that there is no simple cause/effect relationship in concerning technology innovation and diffusion. Gold prefers to conceptualize the relationship as a "matrix of interactions" as he writes:

> "Most such innovations [that is, major technological innovations] in industry do not produce clearly defined, fixed, and measurable shifts in a set of industry-wide economic relationships. On the contrary, most such innovations trigger a continuing process of change within an intricately integrated system of physical resource flows and associated economic valuations. It is through the ensuing complex of interactions that innovation is 'digested' by progressively more far-reaching adaptations and its effects thereby diffused through the system over time. As a result of such dispersion of successive impacts, and the offsetting reactions they engender, the distinctive effects of a given innovation become

> increasingly indistinguishable. Hence, the objective of determining the effects of major innovations usually has to be reformulated . . . in favor of tracing the time path of the successive changes through which major technological innovations are gradually absorbed into the network of economic relationships."

The government's role in technology diffusion has often been of major importance. However, the shifts in power made possible by the ownership and control of new communication technologies has never accrued primarily to the government. The influence of American ideology is mainly responsible for this state of affairs. Freedom is a dominant value in American culture and this translates into a desire for minimal governmental regulation. The case of telegraphy highlighted the value placed on free enterprise, free speech (as it was a communication technology that the government was attempting to gain control of, and hence information) and minimal government interference (and certainly without government ownership). This was a pattern that was repeated for other interpersonal communication technologies. Government versus private ownership became an issue with the telephone and the fax machine as each of these new inventions affected relations of control within society. However, the mandate was always the same. Public sentiment, influenced by ideology, combined with business pressure to keep the government out of business enterprises that were believed to be operating adequately. And, to reiterate Giddens (1991) view, the expansion of modern institutions is related to their ability to control communication media. Hence, the expansion of the dominance of the economic institution appears to be, in part, directly related to its control over information processing. Interestingly, even the computer, whose invention was directly underwritten by the government for military purposes, was innovated using private industry as the vehicle. Furthermore, the government's involvement was with the hardware aspect of computer technology. Software development, the aspect of the technology that controlled information processing, was innovated by and owned by private industry. Hence, the computer industry as it moved from a government decoding device to a business computing machine to a communication medium is also firmly entrenched in private hands.

NOTES

1. At this point in history, ownership and control over an invention was a matter of choice because the "corporate inventor" (Noble 1977) had not yet replaced the individual inventor. Consequently, the inventor of a technology could be brought into government service and develop it in concert with governmental goals, or he could be left free to develop it commercially, on his own. For a full discussion of the implications of the importance of ownership for the development of technology, see Chapter 6, the Telegraph and the Legal System—patent law.

2. The importance of communication for the military appears to have been more obvious to the French and other European countries as they were frequently at war with their neighbors. This was clearly demonstrated by Halzmann & Pehrson (1994) in their article on the optical telegraph, the predecessor of electro-magnetic telegraph. Claude Chappe invented the visual telegraph and sought support for its development. In a speech given before the French Assembly, Deputy Charles-Gilbert Romme, strongly supported government financing and control of Chappe's invention. According to Halzmann & Pehrson "He gave a speech that strongly supported Chappe's work, emphasizing the potential of the invention for military purposes." The first message transmitted over the official French system in 1794, announced "the recapture of the city of Le Quesnoy from the Austrians and Prussians, . . ." So it came as no surprise that the electric telegraph which increased the ability to control information because it wasn't limited by time-of-day or weather conditions, should also be coopted by the government.

3. A detailed discussion of the various bills presented before congress between 1866 and 1874 can be found in Lindley's (1975) book, *The Constitution Faces Technology*. However, for the purpose of the present discussion, the bills presented in 1869 can be considered representative. During the third session of the fortieth Congress, 1869, three bills were presented to the House of Representatives for the construction and operation of telegraph lines, which would, if passed, have led to the take-over of the telegraph industry by the government. Mr. Washburne, of Illinois, introduced the first bill which called for the construction of a telegraph line with four wires between Washington, D.C. and New York which was to be operated by employees of the Post Office. The second bill would have allowed Congress to grant permission for the incorporation of the United States Postal Telegraph Company under the direction of Mr. G. G. Hubbard and his associates. It would have given this company the right and the ability to establish telegraph offices in connection with the Post Office from coast to coast. It would have

taken away the right of private companies to compete with the government telegraph (with the government buying up the existing companies at a "just" price). The third bill effectively said that the government should have the rights to all lines of communication whether by post-route or otherwise, and, that the government should fix the cost of messages and tariff and duty rates. Additionally, I have taken the liberty of encapsulating the arguments of Western Union and the press over several years into one argument for the sake of simplicity and clarity.

4. The sources for this summary are House Miscellaneous Document #73 of the 42nd Congress, 3rd Session, House Executive Document #35 of the 40th Congress, 3rd Session, The House Committee on Appropriations, April, 1871, House of Representatives Bill #1083.

5. From the internal Western Union files entitled "Quotation from A.T.&T. Co. Annual Report for the year 1918", p.1.

6. Ibid, p.1.

7. The government was instrumental in computer technology prior to, during and after World War II. Though the government did not attempt to gain ownership of this new technology, it did harness its ability in the name of national security. The next section of this work discusses the role of the government in computer technology in depth.

V
The Telegraph and Culture

Technology is affected by culture in as much as there are various cultural factors which influence the process of innovation. Mokyr (1990) tells us that the process of invention and adoption are influenced by: life expectancy, short lives mean little time or incentive for innovative activity; a society's attitude toward risk taking for people in various positions within the social structure; geographical environment; prior technology, or level of development; religion; war; demography; and politics. The level and kind of inventive efforts are circumscribed by cultural values and norms. Reciprocally, new technologies that are successfully diffused throughout a society will impact on these same values and norms, thus mediating the experiences of social interaction. This chapter explores some of the areas in which telegraphy affected the social fabric of American society. It also illustrates the extent to which telegraphic technology became diffused into American culture, becoming more and more embedded with each new use. Some of the cultural effects of telegraphy presented here were of real importance, others reflect more creative uses, but all point to the degree to which telegraphy captured the hearts and minds of American society. Whether the new uses were whimsical or serious, all were expressions of the embedded idea, if not the direct physical application, of what telegraphy stood for: speed and instantaneous communication. This chapter offers a brief survey of some of the cultural effects of the invention of telegraphy.

TIME

Time is, in part, a social construction. Each culture breaks down time into units that are useful and meaningful to it. For example, American society is familiar with the ideas of a "work week" and a "school year"[1], both of which are organizing concepts for daily life. However, these same ideas would be essentially meaningless to the primitive Bushmen of Africa as they have no relationship to the circumstances under which they live.

Throughout history, people's relationship with, and conception of time has changed. Quite often it is technology that mediates our experience with time. Telegraphic technology can be thought of as the first communication technology to drastically alter our conception of, our relationship with, and our control over time, setting the stage for our modern preoccupation with lack of time.

Prior to the invention of telegraphy, time in the United States was not standardized. There was a sense that control over how time was spent during the course of a day had human agency. Time and space distanciation were clearer, that is, the distinction between past, present and future and between here and there was more concrete. Time was something to be managed; it had not yet been quite commodified.

Other technologies contributed to the quickening of the pace of life in America such as the introduction of the railroad and mechanization, altering people's relationship with time. However, the application of telegraphic technology to time had the effect of standardizing time, making it subject to more precise management (thus making it a tool of corporate management in the interest of gaining profit through better organization and use of time), and, turning it into a commodity controlled by experts and offered for sale.

The Western Union Telegraph Company, in conjunction with the federal government, was responsible for applying telegraphic technology to time, standardizing it and disseminating it across the entire United States.

The "Time Service" branch of Western Union was a cooperative effort between Western Union and the U.S. government. The arrangement worked as follows: Each day government *experts* at the United States Naval Observatories in Washington, D.C. and Mare

Island, California, would determine, through astronomical observations, the correct mean and solar time. The results of these observations were recorded by means of a chronograph, and from which the transmitting clocks were set and regulated. These clocks were connected by wire with Western Union's operating departments in Washington, D.C. and San Francisco. Beginning at 11:55 a.m. and ending at exactly 12 noon at the 75th and 120th meridians, the time shown by the Naval Observatory transmitting clocks was transmitted in seconds to the operating department where beats were picked up and automatically transmitted over the entire company's telegraph system throughout the United States. This became the standard time of the United States and this time used in the operation of every railroad system across the country. This standard time was also adopted by businesses across the United States. For this particular aspect of the Time Service, Western Union received no direct compensation. However, Western Union managed to commodify time by gaining revenue from the rental of self-winding, synchronized clocks, and regulating these by hourly signals of time correction. Western Union also garnered revenue from the sale of its regulating service to people who owned, outright, self-winding clocks.[2] While the establishment of standard time was probably not of major importance economically as far as Western Union was concerned, it was of greater importance symbolically, as it aided in the process of nation building by establishing another level of commonality.

Arguably, the standardization of time can be viewed from polar positions, either 'time' became controlling or people gained control over time through standardization. The ability to schedule more precisely meant "time", in a sense, was more in control, dictating people's actions, rather than people dictating action. For example, prior to the standardization of time, and the application of telegraphy to railroad technology, people often spent a lot of time waiting for trains at depots. Subsequently telegraphic technology afforded the railroads the ability to control the movements of trains and be more precise in their scheduling. This meant that the pace of travel by rail quickened, less time was spent waiting for trains, and the train depot as a place where people socialized, that is, exchanged general conversation, gossip and news slowly disappeared. On the other hand, people gained control over time by not having to waste time hanging around the depot for extended periods of time.

The tempo of American life was changed by telegraphy. Instantaneous communication meant there was less time for decision-making. Responses had to be more immediate. The pressure of time was "on". Americans, always aware of time, became even more conscious of concepts such as "saving time", having no time", "running out of time" and being "up with the times". An example of each of the foregoing will help to illustrate the point. On: saving time: A pamphlet entitled "Suggestions for Social Messages via Postal Telegraph", published by Western Union, instructs "Its [the pamphlet's] purpose is merely to assist you— . . . to save your time and effort when it comes to phrasing your words for suitability in telegrams." On: running out of time: Further on it suggests that a social telegram is the most expedient in cases where " . . . you may forget or postpone till the last minute to take care of these obligations." On: having no time: An actual telegram from the Western Union Archives dated December 24, 1880 and sent from D. B. Platt in Norfolk, Virginia to Harry C. Platt in Washington, D.C. read "All send Merry Christmas to you all. Too busy to write. Sent box express yesterday." On: keeping up with the times: Western Union telegraph stamp instructs "Be modern—telegraph." With each successive advance in communication technology, the pressure associated with timeliness increased. Around the turn of the century, the availability of telephones in homes as well as businesses meant that people became more accessible day and night. Computer, beeper, cell phone, and fax technology are further expanding the scope, the ability, the necessity of instantaneous communication to the point where there is often no clear distinction between work time and leisure time. From Benjamin Franklin's adage that time is money, to the oft heard expression today that timing is everything, we can gain an intuitive understanding of the increasing prominence of "time" as a social construct in modern society.

HABIT OF COMMUNICATION

Bellah (1985) borrowed a phrase from de Tocqueville, "habits of the heart" to describe concepts that have become an integral part of American life, even though they may cause conflict and produce stress at times. Belief in these concepts are deeply ingrained in the American psyche. The idea of access to instantaneous communication can be thought in modern society as a "habit of the heart", one that began with telegraphic technology. E. B. White (1939: 13) writes of the telegraph "What had once been considered an impractical invention the Government had refused to buy for $100,000.00, had by 1889, in private hands, become indispensable to social and economic life." The swift growth of the American District Telegraph Company is also testament to the intensity with which Americans "took" to instantaneous communication. The American District Telegraph Company was established in 1872 with 4 subscribers. It offered telegraphic capabilities from within the convenience of the home for a limited number of services, including the ability to instantly summon a messenger to attend to one's needs.[3] In less than 2 years, 12 additional offices were open and the company had over 2,000 subscribers. One could say that these devices paved the way for future technological advances in communication technology to be adopted for home use.

TELEGRAPH TECHNOLOGY AND MEDICAL PRACTICE

A less widely known application of telegraph technology was its use in medical practice. The connection between communication technologies and medical practice, particularly as it relates to diagnosis, which seems like a recent development, was actually already present in the age of telegraphy.

Dr. William Channing, a medical doctor, was intimately involved with bringing about the practical application of telegraphy to the fire alarm system. It appears that an interest in the practice of medicine and an interest in communication technology were not incompatible. In fact, one way of conceiving of the practice of medicine is as a relationship between the doctor and a human body. Relying solely on the patient's verbal communication of what he/she is feeling is not always accurate or possible. Hence, it is a somewhat inefficient means of communicating with the body, one relying on a "middle-man" so to speak. Doctors have looked for more precise means of communicating with the body to aid in their diagnosis and treatment. Aronson (1977: 71) tells us about the role that the telephone played in the field of health care as referenced in the *Lancet* between the years 1876-1975. One of the ideas he discusses is the application of telephone technology to aid in medical diagnosis. One such account states:

> "During the early days of the telephone there were other references in the *Lancet* to ways in which the new instrument can aid diagnosis. One device which was developed in Great Britain and the United States was the 'telephone probe' used for locating bullets and other metallic matter lodged in the human body. Indeed, Alexander Graham Bell himself was the inventor of the telephone probe and had been called to Washington in 1881 by doctors trying the locate the assassin's bullet that had felled President James Garfield."

However, telephone technology was not the first communication technology to be used in the delivery of health care. A *Harper's New Monthly Magazine* (1873: 359) article on the telegraph reported a remarkably similar story. "One of the most curious applications of the telegraph is its use in surgery to discover a bullet would. The probes and forceps are each connected with a delicate battery. When one point of the probe or forceps touches the ball no effect is produced, but when both touch it the ball completes the circuit, and the tinkling of a bell or the vibration of a spring shows the surgeon he has seized it." Clearly, it isn't too much of a leap of faith to assume that Bell's invention was informed by earlier work in telegraphy. Admittedly these were novel and somewhat indirect applications inspired by telegraphy and telephony.

Telegraph technology was also popular as a consultative tool. Lardner (1857: 91) informs us that "Nothing is more frequent in the United States than electric medical consultations. A patient in or near a country village desires to consult a leading medical practitioner in a chief city, such as New York or Philadelphia, at four or five hundred miles distant. With the aid of a local apothecary, or without it, he draws up a short statement of his case, sends it along the wires, and in an hour or two receives advice he seeks, and a prescription." More recently, fax and computer technology have been used to aid in medical diagnosis and treatment. Both technologies can make a patient's medical history instantly available, as well as allow for comprehensive consultation through shared databases and communication networks.

SOCIAL TELEGRAMS

Responsibility for changing one of the norms of 19th century society can be placed in part with the telegraph. Stern (1988: 6) asserts that history indicates that the first card-worthy occasion was the New Year and that when a gift was delivered from one sun-baked brick bungalow to another, a message went with it. "In such a manner—written in papyrus or uttered by the messenger—were the world's first greeting bestowed. The Romans followed with copper pennies showing the two-faced god, Janus, then pictured with greetings on terra cotta and medals. The next known example, a woodcut commemorating the New Year, appear in Germany in 1450." And, prior to the advent of telegraphy, it was considered proper for people to visit each other in person, on holidays or for special occasions, such as birthdays, weddings, graduations, Christmas, Easter and New Year's. In the absence of such personal visits, hand written notes and letters were considered acceptable form. Initially restricted to business use, use in emergencies, or for the transmission of sad messages (bad news), telegrams gradually and steadily gained acceptance for social correspondence. White (1939: 16) writes "Thus we find that social telegrams, . . . have in recent years won such widespread acceptance that social arbitrators now recognize telegrams as correct for all types of formal, as well as informal correspondence." (It is important to note the class influence in this instance —"social arbitrators" implies that it was the people of the dominant class that had the power to change norms.) Beniger (1986: 274) in discussing the control of consumption, tells us that mass media helped secularize Christmas and other festivals. For example, "By 1873 the illustrated periodical press promoted Easter as an occasion to display the latest fashions in men's and women's clothing." *Harper's Weekly*, (April 26, 1873). Holidays moved from the realm of private affairs to commercial affairs, and such a move may have contributed to the depersonalization of these occasions. Further evidence that the public had accepted the sending of impersonal greetings for special occasions was the introduction of the special "holiday blank" (a telegram form embellished with decoration for the appropriate occasion) by Western Union in 1912.[4] Shortly thereafter, additional decorated blanks became available for other occasions, followed by prepared texts for those who needed help in find the right words for the right occasion. Social telegrams, as an acceptable form of inter-personal communication caught on and became a profitable area of business for Western Union. In 1934, in order to increase the use of social telegrams, Western Union instituted a "flat rate" for the transmission of standardized, numbered messages that were suitable for weddings, birthdays, the New Year and many other occasions. In order to facilitate the public use of social telegrams, Western Union issued franks and advertising stamps "Good only for Family and Social Messages. Not to be used for business or political communication." This standardization of "personal" greetings can be seen as one additional step in the process of homogenization.

LITERATURE AND LANGUAGE

One indirect way of measuring the diffusion of a new technology is to examine its impact on the language of a society, and its incorporation into the literature of a culture. Language and literature are both reflective of a society's norms and values, and will yield a clue as to how deeply embedded a new technology has become.

America was enthralled by and romanticized the use of the telegraph. Telegraph tales involving heartfelt and sometimes unusual sentiments permeated the literature of the late 1800's. The mention of or reference to the telegraph could be found in many forms of literature. Johnston (1882) compiled a collection of such literature which deals with a variety of themes from love to burglary to intrigue. Examples: "Kate, An Electro-mechanical romance.": 53; poem—"Song of the Wire" p. 144; play—"The Carnival of Oshkosh" pp. 177-189; song- "The Telegrapher's Song" p. 157 short story—"Into the Jaws of Death". If literature mirrors society, then one can clearly see how intrigued society was with this communication technology and how integral a part of life it had become.

New words were introduced into the English language. For example, "telegram" came to replace the phrase telegraphic dispatch or telegraph communication. "Teletype" was developed to refer to any printed as opposed to written message. "Telegraphery (an office) and "telegraphage" (rate for sending) were also added between 1848-1847, but did not catch on. Coates (1976) Additionally, the word telegraph came to be associated with anything speedy. For example, John E. Fuller decided to name the device he invented for performing mathematical calculations, the "Computing Telegraph". A notice in the Manchester Guardian of Monday, March 9, 1868 describes the use of this name as follows: "The name is a novel application of the word 'telegraph', with no relation to the derivation, and obtaining its significance from the speed which is characteristic alike of this computer and of the electric wires".

One final illustration of how captivated the American psyche was with telegraphy can be found in the somewhat exaggerated assertion that the telegraph brought about a standardization of communication through the introduction of the new language, Morse code. An editorial in *Telegraph and Telephone Age* (June 1, 1932) expresses the opinion that a new language, one truly international, came into being with Morse code. Those who have studied the science of written languages tell us that the Morse code was the first new alphabet to be created since the ancient Phoenicians drafted their linguistic symbols. "Communications between people was old centuries before Morse was born, but for all those centuries no real progress in the art had been made, either in means or methods. Some acceleration there had been, or course, but none of actual consequence. Then Morse, at one stroke, broke through space and pointed the way to what the world enjoys today and to the greater facilities it will have tomorrow."

Speech pattern and tones can be indicative of class status. Use of Morse Code as a standardized structure for the language of telegraphic dispatches meant that telegrams, in one sense, were a classless language. For example, A *Harper's New Monthly Magazine* (1873) expands on this idea when it states "And as a whole it may be said that the science of language in the hands of philologists is used to perpetuate the differences and irregularities of speech which prevail. The telegraph is silently introducing a new laments, which, we may confidently predict, will one day present this subject in a different aspect." This article also states that Morse's invention gave preeminence to the Italian (Roman) alphabet and thus assured its adoption, or some improvement upon it throughout the world, due to the necessary convertibility of expression between different languages. The article asserts that it was the press, via the influence of the telegraph, that was impacting on language, creating a certain amount of assimilation. For example, an even transpiring in any part of the world is concisely expressed in a dispatch, which is immediately reproduced in a number of different languages, and, as the *Harper's* article states "A comparison of such dispatches with each other will show that in them the peculiar and local idioms of each language are to a large extent discarded. The process [telegraphy] sifts out, as it were, the characteristic peculiarities of each language, and it may be confidently said that nowhere in literature will be found a more remarkable parallelism of structure, and even or word forms, combined with equal purity and strength in each language, then in the telegraphic column of the leading dailies of the capitals of Europe and America." Short words gained an advantage over long words, words of precise meaning gained advantage over ambiguous ones, direct forms of expression gained an advantage over indirect ones, and local idioms became a disadvantage everywhere. To the optimists of the time (or at least to the writer of this article) the idea of a common language of the world seemed within the realm of possibility "And it is significant of the spirit of the times that this idea, once so chimerical, should at the time we are writing find expression in the inaugural of our Chief Magistrate, in his declaration of the belief 'that our Great Maker is preparing the world in His own good time to become one nation, speaking, one language, and when armies and navies will no longer be required.'" (p.360) This false dream may have been an expression of the values upon which America was founded. America represented the land of opportunity, equality and justice for all. This implies a belief that diverse people could live together harmoniously. It reflects a naive hope that better communication would yield more accurate knowledge and that this would lead people to emphasize commonalties rather than differences. However, such dreams ignore the realities of the needs of the various economic institutions within society and the ways these institutions' needs are reflected in the values and norms of other social institutions. Though this prophecy has not come true in its exact form, there is a sense that people still look to communication technology to play a role in creating the global community. (See Chapter 8 for a fuller account of this idea.)

An interpersonal communication technology that is successfully diffused embeds itself in a given culture altering relations within that society. The overt effect of telegraphy was the way it was able to mediate the effects of distance in regards to communication. Often there are other tangible and intangible effects on relationships. Intangible, but important, are the psychological effects of a technology as they can alter outlook and perception such as exemplified by the standardization of time, or the development of the habit of instantaneous communication, which has come to represent more than just the ability to communicate (for example, it represents security, sociability). This habit continued to grow and continues today with access to the Internet. Additionally, the appearance of a technology in the literature of a society reflects the ability of the invention to capture the creative imagination of a society.

Tangible applications have a more direct impact on culture, such as the addition of new words to the language, the changing of norms as exemplified by the increasing use of impersonal greetings (which then made possible the establishment of the commercial greeting card industry), the temporary leveling of status differences by the use of temporary employees (the messengers), or the ability to have access to medical consultations regardless of the distance between patient and doctor. While these are all examples of the cultural impact of telegraphy, we have seen that other subsequent innovations in interpersonal communication technologies impact on some of the same aspects of interpersonal relations. For example, the ability to consult with doctors without having to be in close physical proximity has been advanced by the telephone, the fax machine and the personal computer via its modem capabilities. The telephone has become an acceptable means by which to offer greetings for various occasions, and E-mail has the ability to replace formal impersonal greetings (cards) with more personal, direct computer mediated greetings.

NOTES

1. Eviatar Zerubavel takes a sociological look at time as it relates to daily life in his books *The Seven Day Circle: the history and meaning of the week* and *Hidden Rhythms: schedules and calendars in social life*.

2. These self-winding clocks were connected by wire to master clocks in the company's in the Western Union's time service departments. These master clocks are the ones regulated from National Observatory signals, and are then used to automatically correct the clocks connected to its circuit.

3. Early on in this history of telegraphy, messengers were used for a variety of purposes other than the delivery of messages. Women often found messengers useful as escorts (to and from places of amusement, from one house to another in the evenings, to walk children to/from school); to baby-sit, to do shopping; to pay bills; pay New Year's calls by proxy. A *Brooklyn Daily Eagle* report on March 9, 1910 states. "A unique service furnished by the local company [Brooklyn Telegraph & Messenger Service] is a page service. Uniformed pages are furnished to society matrons for special occasions. During the social season, demand for the company's pages are heavy." (p.60) Messengers were also often used to distribute political documents. The availability of the telegraph messenger for services other than the delivery of messages introduced a new type of temporary employee into society. The use of these messengers mediated the effects of social status for some middle class families by affording them access to lifestyle services that the wealthier members of society have on a permanent basis.

4. Interestingly, this coincided with the official establishment of Hallmark, the greeting card company. Social telegrams and pre-printed messages were further institutionalized into the commercial greeting card industry. It is interesting to note that consumers now have a choice, via technology, to personalize these impersonal cards, instantaneously.

VI
The Telegraph and Social Stratification

Stratification, in its essence, is descriptive of a relationship of inequality. There are various forms of stratification within society, but two forms, social class and gender, are particularly germane to this work. The impact a given technology has on a given culture varies by relative position within the social hierarchy. Gabler (1988), in compiling a social history of telegraphy in America between 1860 and 1900, instructs us that telegraph technology facilitated the development of a new social class within society. The appearance of the new "lower middle" class is explored in this chapter. The second part of this chapter looks at the way women experienced telegraphy in terms of both their class and gender.

THE EMERGENCE OF THE NEW LOWER MIDDLE CLASS

Nineteenth century America was a country in transition. The economic base was changing from one based on agriculture to one based on manufacture. Consequently, the structure of society was affected by this shift. The center of work moved from the home to the workplace, changing both the organization and structure of work. A clear distinction was made between the domestic sphere and the public sphere, with women being relegated to the domestic sphere. A change in the guiding ideology also had to take place to institutionalize the newly emerging roles within industrializing America. Kessler-Harris (1982) captures this change when she writes " . . . [T]he old Puritan ethic which stressed morality, hard work and common welfare, was supplemented by the ethic of laissez-faire economics, which emphasized individualism, success and competition." Family based collegial networks were replaced by impersonal, corporate, competitive ones.

The newly emerging corporate structure meant a re-structuring of work organization. The introduction of the bureaucratically organized, hierarchical firm offered new employment positions and opportunities. Telegraphy offered employment for men that was above blue collar work, manual work, and something less than the entrepreneurial work of the traditional middle class. Telegraphers, as Gabler (1988: 57) informs us, " . . . were among the very first mass of white collar employees". Gabler further instructs us that during the transition period of the nineteenth century, the "old" middle class was giving way to a "new" middle class, and that telegraphers were among the lower echelons of this new middle class.[1] The fact that telegraphy was a national phenomenon, with standardized work practices and salaries, made telegraphy an important vehicle for the emergence of this new level of the middle class. Telegraphy was an industry large enough in both scale and scope to foster a common national identity and culture among its telegraph operators. Much of this newly emerging culture was disseminated through and reinforced by union membership. There was a national telegraphers union with local branches throughout the country. Also, journals and other professional associations helped to create and foster this common identity among telegraph operators.

There were also other status symbols that accompanied this new position within society. Telegraphers could distinguish themselves by virtue of their education, as they usually possessed an education above that of the average working class laborer. At a time when most people were laborers, farmers, or craftsman, telegraph operators had a different standard of dress, they wore gentleman's attire, suits with collars and cuffs. Their work was also clean and perceived as genteel. Telegraphic employment was clearly a mechanism for social mobility. Gabler (1988: 59) tells us that among telegraphers in big towns and cities, 64% were the children of blue collar families.

Lest one get too carried away with the way telegraphy appeared to mediate the effects of class, it should be mentioned that this new job position was actually a double-edged sword. On the one hand, it increased social standing by drawing people from the working class up into the middle class (even if it was at the lowest level). However, the other side of the coin has to do with the employees status within the work culture. The bureaucratic, hierarchical structure of Western Union meant that the workplace was stratified as well. Telegraph operators were among the lowest level employees, sharing their place with clerks and bookkeepers for example, people with a skill and usually no more than a high school education. In contrast, Gabler (1988: 68) informs us that Western Union Electricians (or Electrical Engineers, to offer a modern job description) were usually college educated and from the upper middle class.

Obviously, it was not telegraph technology alone that led to the emergence of this new social class, but it is clear that the infrastructure associated with the emergence of a national telegraph industry, certainly played the role of facilitator.

OCCUPATIONAL SEX-TYPING

Gender stratification ensures a hierarchy in which men have higher status than women, and this inequality leads to domination. Cockburn (1985: 251) writes "Gendering appears to have been, for as long as we can see backwards in time, a major organizing principle, if not 'the' organizing principle, in our perception of the world and everything in it." One of the areas in which her words can most clearly be illustrated is in the workplace.

As stated previously, one of the major features of the ideology of laissez-faire capitalism is competition. One way to restrict the amount of competition for jobs is to ideologically disqualify a whole category of people from work. This is what Victorian ideology effectively accomplished. This ideology touted that women's place was in the home (the domestic sphere). It was an ideology that defined women primarily in terms of their biology, and not at all in terms of their ability or individuality. Historically, women's work was framed in terms of the family's needs or society's needs, not in terms of personal satisfaction or expression. This ideology mediated women's experience with work in the nineteenth century, and for generations afterward.

The ideology of the Victorian era in which telegraphy developed clearly dictated that "ladies" did not work, and those women who did work (factory laborers and domestic servants) were essentially invisible to society, i.e. they were of a social class that could basically be ignored, owing to the duality of the conception of women's place within Victorian society. The women of the Victorian Era that were the ideal of the time were the upper middle class and upper class women. It was to these women that the ideology of the time was directed. (In other words, it was these women that the economic system wanted to inhibit from participating in the work force.) Hence, middle class "ladies" did not pursue telegraph jobs because of social prescriptions regarding their place in society

Technology can also be conceived of as a medium of power, developed and used in any system of dominance to further the interests of those at the top. Historically, men have controlled technology—its knowledge, development, and application. This means that they have also acted as the gatekeepers of technology, deciding when, where and to whom that technology gets diffused. Women's relationship to technology has generally been second hand, and telegraphy was no different, especially as it mediated women's experiences of work.

Telegraphy was sex-typed as male. Cohn (1985: 12) instructs us that sex-typing of occupations is affected by prevailing rates of female labor force participation at the time at which a give occupation is created. "If the initial job openings occur in a period when traditional sex-role ideologies are inhibiting female labor-force participation, no women will apply for the new vacancies and the occupation will become male. If the job is created in a period of increasing female labor force participation, it will be possible to staff the positions with women." Working class women were initially excluded because they did not have the resources or education necessary to pursue the job. At the inception of telegraphy, what meager resources that were available in a working class family to acquire a skill or education, were usually spent on the male offspring. As a way to further discourage women, telegraphy was depicted to be physically demanding as well as potentially dangerous in that the first telegraph operators worked lengthy hours (average of about ten hours a day), performed other tasks such as wire maintenance which often required that they climb poles and handle tools, and that they often worked in sparsely populated and somewhat isolated areas. Telegraph work was depicted as being entirely too dangerous for women.

However, some pioneering and socially active women did apply for the job of telegraph operation. Sarah J. Bagley, a woman activist in the area of women's labor in the mill industry, became the first female operator in the Western Union Telegraph Company on February 21, 1846. As operator and manager of the Lowell, Massachusetts office she proved that women were capable or participating in this new technology as workers and not just users. More women were to follow in her footsteps. An editorial in *The National Telegraphic Review and Operators Companion* (April, 1853) noted "Lady telegraphers are getting more numerous." However, women's place was not yet secured. This same writer, reflecting the view of his time, continued "One good woman on a line is a treasure, a sure means of preserving the courteousness which woman's presence so universally encourages and secures." But the writer also felt that by nature, women could not "mend a break, buckle on climbing spurs and ascend slippery poles." He—for surely the writer was male—had not heard of the female operator who climbed a telegraph pole during a storm, repairing the line and thereby saving a train from crashing. Brady (1899: 32) Still, the perception was that women's place was in the home. Brady's writing also testifies to this when he tells us that this same heroine telegrapher, two months afterwards, " . . . married the chief dispatcher, and the profession lost the best woman operator in the business."

Feminization of Telegraphy

The Civil War was the catalyst for the increased opportunities for women in the telegraph profession. Cohn (1985) indicates that one means for women to break into a male industry is when there is some sort of limit on the acceptable male candidates. The great demand for soldiers during the Civil War meant that there was a shortage of male labor. Juvenile males (boy telegraphers of 14 were not uncommon during the Civil War) were often engaged as telegraphers by the Military Telegraph to fill the void. Consequently, women often took up the posts in the commercial telegraph centers. However, the war alone was not responsible for opening up telegraphy to women. For example, Western Union had become a national company by 1861. Increasing size and complexity meant that there were structural changes within the organization that impacted on employment decisions. The number of levels in the organizational hierarchy increased, opening up the possibility of promotion to operators. Men involved in telegraphy had a career ladder to climb. Additionally, innovations in technology meant that the job of operator became stratified. Operating the single circuit lines was the least complex and hence the lowest position for an operator. Operating the multi-circuit lines was the most complex, required greater skill and hence was accorded the highest position among operators, that of First Class Operators. To a certain extent, one could say that the concept of de-skilling was at work. Learning simple, single line telegraphy was no longer perceived as a real skill; real skill meant having the ability to operate the multiplex, heavy traffic lines. Women telegraphers were habitually assigned to the easier wires—that is, the single circuit, light traffic lines, and usually segregated from the men in what was known as the Ladies Department, further isolating them from the more skilled work. It precluded them from being able to pick up the skill by filling in during the less busy times for the more skilled operators. Consequently, even when women did become operators, they had inferior skills by design in as much as they were confined to these positions. Ideology played a role in justifying this practice. Women workers were viewed as temporary workers (until they married), hence there was no need to promote them or put them on a career track. Leaving this niche primarily for women also had economic benefits—the least skilled jobs received less pay and women in these positions received less than their male counterparts, even though they worked the same number of hours per shift. (This practice too, was justified via ideology—women were not supporting families therefore they did not need a "family wage" they could do with less.) For example, the Bureau of the *Census Special Reports on Telephones and Telegraphs* (1902) has a section on salaries and wages. From this report we find evidence of the gender wage gap: the average number of male telegraph operators was 10,179, and their annual wages for the year totaled $7,494,909.00, an average of about $750.00 per year per operator. Female operators numbered 2,914. Their total wages for the year was $1,367,440.00, an average of about $440.00 per year per operator. This represents a difference of about 59%. It is interesting to note how amazingly similar is the rate differential to that which we have today (about 67%). One is led to wonder why women took these jobs. One explanation was that it was in answer to a social problem. Johnston (1882: 50) tells us about a writer, W. J. Foster, who was interested in the question of why women became telegraphers. Foster wrote "[Telegraphy] gives employment to women as well as men, and this assists in the practical solution of the difficult question: What must society do with the capable and intelligent female population who cannot marry, for the very sufficient reason, among others, that there are not enough men to mate every one of them?" An alternative explanation had to do with opportunities available to women at the time, and demographics. Although women's wages were lower than men's, telegraphy was still a job that paid better than anything else available to women and, it was a more socially prestigious job than others open to them (i.e. primarily factory work and domestic service). Consequently, the job of telegraph operator was more desirable. Secondly, Gable (1988) informs us that demographics had much to do with the abundant supply of women. The great famine of the 1840's meant that Irish immigration to the United States was great[2], and unmarried women in particular, swelled their numbers. Gabler (1988: 119) tells us "that among a sample of 70 women telegraphers in New York City in 1880 with at least one immigrant parent, nearly three-fourths of them—71.4%—turned out to be Irish." And, the *Census Report* of 1900 reports that a plurality of female telegraph operators (41%) were of Irish extraction. Consequently, there was a large pool of unmarried women from which to draw labor. (It should also be mentioned that casualties of the Civil War also contributed to the ranks of unmarried women.)

It could also be hypothesized that the principle of synthetic turnover was also at work in this feminization process. It could be seen as a solution to an economic problem, that is, corporate profitability. An article in the *Journal of the Telegraph* (January 15, 1869) states

"With the pressure constantly brought to bear on telegraph companies to cheaper rates, . . . we expect to see demands made for women to serve in telegraph offices for beyond what now exists . . . for short lines, for office work, for a variety of work yet to be introduced into the labor of telegraph offices, . . . it is easy to see that the service of woman, must, in the very nature of things, be largely demanded."[3] This statement reflects more than just a writers opinion on the state of telegraphy. It represents the thinking of those eminently involved with the industry. The new administrative structure of Western Union was profit oriented and driven. Gabler (1988), instructs us that in order to achieve economies, it was the practice of Superintendents to fill any vacancy in an office at a lower rate than the existing occupant received. This meant that women, with their inferior skills, limited work life-span, and abundant availability, were prime candidates for these positions. Additionally, around 1868, Western Union began cutting salaries as corporate concentration continued. At the same time, Western Union was also aware of the increasing unionization of its labor force, and the growing disenchantment of these unions to the wage cuts and increasingly austere economic policies being enacted by Western Union. In response to these economic realities, an in conjunction with Western Union, Peter Cooper, an industrialist who was also involved in the laying of the transatlantic cable, opened a school of telegraphy especially for women at the Cooper Institute in New York City in 1869. This proved to be a particularly propitious move for Western Union as the training program started only one year before the strike against Western Union in 1870. Western Union was in a good position to weather this strike because it had a large pool of operator labor with which to replace the strikers. Gabler (1988: 150) tells us that the strike lasted only one week. "League weakness as much as Western Union's strength accounted for the quick collapse."

To the extent that the feminization of telegraphy took place (and the statistics of women's employment in telegraphy does show an increasing trend—see Table 6-1) one can say that it was the result of evolution, rather than planned, as was the British feminization of telegraphy, which was far more complete because the management of the Post Office (which controlled the telegraph industry in England) committed itself early on to a female signaling forces in order to economize on personnel.

Table 6-1 Women Employed in Telegraphy

Year	Category	Number (% of total)
1870	Non-clerical telegraph employees	355 (4%)
1880	Officials and employees	1,131 (5%)
1890*	Telegraph and telephone operators	8,474 (16%)
1900	Telegraph operators	7,299 (12%)

Source: Census Reports of 1870-1900.

*Note: The census of 1890 combined telephone operators, who were already heavily female, with telegraph operators so the number of women employed in these industries rose radically. However, when telegraph operators were separated out again in the Census of 1900, it is evident that there is a trend for the increase of women in the telegraph industry.

Telegraphic technology not only made possible the movement of women into this new social class, but basically opened the world of corporate America to women. Limited though their role was, telegraphic employment afforded a woman the opportunity to pierce the barrier between the private and the public sphere without jeopardizing her position as a lady. Additionally, telegraphy paved the way for the role of women in the next innovation in communication technology—the telephone. By the time telephony became widely adopted, women's employment with "talking machines" was well established. Reflecting a popular stereotype concerning women's tendency to talk, an April, 1853 editorial in the *National Telegraphic Review and Operator's Companion*, (p. 50) asserted that "It is an apparently appropriate occupation for them [women]; the telegraph is a talking machine—who understands that business better than they?" As the system grew and tasks were divided and routinized, it came to be viewed as socially acceptable for women to engage in this low level corporate work. The public sentiment now was that telegraph work was actually similar to domestic work. It was performed indoors, it was not too stressful (women could sit while they worked), it was genteel and civil. If a woman had to work to help support her family, or if her family needed to do something with her until they could marry her off, telegraphy came to be viewed as an appropriate occupation. It was not accidental that this change in social attitude corresponded to the economic needs of business, for as Noble (1977) instructs us, the interests and needs of the dominant institution in a society will stamp the new technology, giving it direction and, by extension, influence the social fabric in order to keep it in sync with those interests and needs. Computer technology illustrates the same pattern, it was sex-typed male. Men were the designers, builders, analysts, and programmers of the computer systems; they had control. Women encountered computer technology second-hand, that is, as users, in jobs that were typically female-related, for example, as data entry clerks, and as word processors. In relation to computer technology, these were the jobs which were at the bottom of the status hierarchy. This is another example of the control and power exercised by the dominant group (men) and mediated through technology which had the effect of reinforcing women's "place" in society. Telegraphy, and subsequent innovations in communication technology, clearly illustrates for us Giddens (1991) concept that mediated experience influences both self-identity and the basic organization of social relations.

NOTES

1. Gabler's use of the terms "old" and "new" middle class follows C. Wright Mills' taxonomy in which a distinction is made between the "old" middle class of businessmen and free professionals, and the "new" middle class, who were managers, technicians and lesser white collar employees—people who would spend their entire lives working for someone else.

2. The Population Census of 1850 shows us that the Irish comprised the largest number of foreign born residents -961,719.

3. That this writer's prediction of the demand for women's service in telegraph offices was based in fact can be demonstrated via history of the evolution of the auditing department with Western Union (discussed in end note in Chapter 3). Women were given the tedious and mundane tasks within this new area.

VII
The Telegraph and the Judicial System

Telegraphic technology insinuated itself into many diverse areas of everyday life from the commercial to the social to the legal. This chapter will offer a glimpse into the telegraph's impact on the legal system. The nature of patent law in the U.S. changed over time, from laws that were meant to protect the individual to mechanisms of control for corporations. This chapter will examine the relationship between telegraph technology and patent law evolution. The second section of this chapter considers the relationship of the telegraph to contract law, and explores the ramifications of intra-industry legal arrangements in the struggle for dominance and control. The section on contract law will also illustrate the struggle for market supremacy among the newly emerging corporate giants. The legal battles between Western Union and the Bell System reflects competition on a scale, and of a scope that was heretofore unfamiliar in American business practice and culture. Business relationships in this era were altering the culture of work even as the new work cultures were impacting on the future of technology innovation via the management and research decisions these firms were making. The final section presents a more novel use of telegraphy in the experience of the criminal justice system. It is included here because it does raise the issue of the relationship between communication technology and what constitutes acceptable evidence.

PATENT LAWS

Patent laws were among the first of a class of laws meant to protect what has now come to be described as "intellectual property" rights, in the interest of encouraging invention and creative activity. Patent laws were originally designed to protect the individual inventor. It was believed that the patent system would operate to benefit both the inventor and society. It was assumed that by protecting the rights of the inventor, he would then willingly share his invention with the public and allow its commercial development (thereby both would mutually benefit). Noble (1977: 87) tells us that "The first U.S. Patent Law, of 1770, was administered by Thomas Jefferson and his colleagues under very strict standards, and relatively few patents were issued. Three years later a more relaxed system was adopted whereby 'anyone who swore to the originality of his invention and paid the stipulated fees could secure a patent,' its validity being decided by the courts. In 1836 this second law was repealed and a Patent Office was created. The 1836 Patent Act 'marked the beginning of our present patent system' based upon the 'examination system' involving scrutiny of each patent application."

Morse applied for his telegraph patent subsequent to the 1836 Patent Act, and the ensuring battle surely impacted not only patent law, but illuminated the primacy of patent rights for the budding capitalist corporations. Thus, Edward Lind Morse (1914: 286), in his biography of his father, Samuel F. B. Morse, writes "The year 1848 was a momentous one to Morse in more ways than one. The first of the historic lawsuits was to be begun at Frankfort, Kentucky,—lawsuits which were not only to establish this inventor's claim, but were to be used as a precedent in all future patent litigation."

The patent battle Morse was involved in revolved around whether or not he was the *inventor* of the telegraph. Morse swore to the originality of his invention though the burden of proof fell on Morse as he did not apply for an American patent in a timely manner. Consequently, Morse had to rely on the testimony of various colleagues and friends to determine when Morse had a working model of his telegraph (which was determined to be around 1835). When news of similar inventions started appearing in journals and newspapers, Morse decided it was time to take steps to legally protect himself. Morse (1914: 69) tells us that on October 3, 1837, Morse sent his finished application (which included a description of his invention) to Washington, D.C. and the receipt was acknowledged on October 6, 1837 by Commissioner Ellsworth (head of the Patent Office). However this move proved to be a little too late to protect Morse from others such as Alfred Vail, who were also claiming that they invented the electromagnetic telegraph. Consequently, Morse found his position as inventor of the electromagnetic telegraph being challenged subsequent to its commercial development. Part of the debate revolved around whether the telegraph was a "discovery" or an "invention". According to Morse (1914: 13) " . . . the word 'discovery' in science is usually applied to the first enunciation of some property of nature till then unrecognized, while 'invention', on the other hand, is the application of this property to practical uses. However, what was at issue here was more than a matter of semantics or honor. Knowing the parentage of the telegraph had commercial implications for the inventor, and ultimately for the corporate monopoly that evolved from telegraph technology. This debate also serves to point out the difficulty with the heroic theories of invention

It was not until January of 1854 that the Supreme Court put an end to all the various patent litigations; "every legal device was used against him; his claims and those of others were sifted to the uttermost . . . The decision of the Supreme Court was unanimous on all the points involving the right of Professor Morse to the claim of being the original inventor of the Electromagnetic Recording Telegraph." Morse (1914: 291) Morse also informs us that his father was almost destitute and had to borrow money in order to survive during the long years and many bills of this patent fight. Subsequent to this favorable decision, Morse was finally able to begin to reap financial reward from his invention. It is no wonder that the argument over the invention of the telegraph raged for 24 years, since money, control and power were all at stake.

The businessmen who were planning the consolidation of the telegraph industry were well tuned to the importance of patent ownership for the empire they were building. The plan of Cyrus Field, et al. was to purchase the patent rights of Morse, Kendall, Vail, and F.

O. J. Smith (stockholder's in Morse's Magnetic Telegraph Company), "and by means of the large capital which would be at their command, fight the pirates who had infringed on the patent, and gradually unite the warring companies into one harmonious concern [Western Union]. A monopoly, if you will, . . . " Morse (1914: 341) As Western Union was expanding and merging with competitors, it was careful to make sure that it bought any patent rights associated with these companies. In gaining control of Morse's patents, and associated patents, Western Union was the first large scale firm to benefit from this corporate control of patents (as well a the first to be hurt by it, as this section will demonstrate).

The example set by the builders of Western Union, as well as the lessons learned from the patent litigations Western Union was involved in, became part of the accumulative cultural knowledge which the builders of the Bell Company system drew upon in creating the telephone monopoly. In this new era of economic restructuring it became clear that expansion and dominance of a business was dependent on patent monopoly. Noble (1977: 11) instructs us that "The life of a patent is seventeen years, after which time the invention becomes public domain. No one understood this fact better than Theodore N. Vail who became the general manager of the National Bell Telephone Company in 1879. From the outset he undertook to occupy the field, as he termed it, so as to ensure that the Bell company would outlive its original patents."[1] This proved to be a critical tactical maneuver. When Alexander Graham Bell offered to sell all his phone patents to Western Union for $100,000.00, the Western Union President Orton, turned him down. Bell managed to gather other backers in the meantime, and had installed a few thousand phones by 1879. Meanwhile, Orton changed his mind about the usefulness of the telephone and bought the patents of another scientist, Elisha Gray. Hounshell (1988: 153) tells us that "In late 1877, Western Union prevailed on Gray to launch a patent fight with Bell . . . Gray filed papers with the U.S. Patent Office on February 14, 1876, documenting his preliminary research in voice communication. Coincidentally, Bell had submitted his telephone patent application on the very same day." These patents were subjected to the intense scrutiny that the revised Patent Law of 1836 demanded. One September 12, 1878 Bell filed suit in Massachusetts, charging an agent of Western Union (Gray) with infringing on his patents. Vail, in his testimony in the Western Union suit hearing explained " . . . how his company proceeded to surround itself with everything that would protect the business, that is the knowledge of the business, all the auxiliary apparatus; a thousand and one little patents and inventions with which to do the business which was necessary, that is what we wanted to control and get possession of." Noble (1977: 11) A long and complicated battle between Bell Telephone and Western Union ensued. Western Union eventually realized that it could not win this suit and the two parties settled out of court. Western turned over to Bell 56,000 phones in 55 cities and agreed to get out of the telephone business for good. A final decree on the case was issued on April 4, 1881, guaranteeing that the Bell Telephone Co. plus its subsidiaries and affiliates would control the new invention.

The foregoing illustrates but two fights (certainly among the earliest) which ultimately ended with control and development of inventions via patent ownership coming to rest primarily with corporations. Noble (1977) tells us that invention became institutionalized; increasingly patents became the property of corporations rather than of the individual inventor. He further states that the solitary inventor had two choices in the face of the development of large scale corporations: fight for their rights and seek to develop their invention on their own, or, form an allegiance with a corporation—selling their patent rights in exchange for employment security, thus also ensuring the financial backing for further research and development. This shift in the relationship between inventor, patent holder, and corporation meant that emerging corporate America gave birth to the idea of the corporation as inventor (as Noble 1977, labels it). Figure 7-1 depicts the trend to labeling the corporation as inventor as evidenced by whom the patent applications were issued to. Even though the groundwork for the change was laid by the first corporate giants (Western Union and The Bell Telephone Co.) by their concern with patent ownership, the invention function did not become institutionalized until the 1900s. By 1901 it had become enough of a phenomenon to warrant separate reporting in the Census Bureau statistics. By examining Figure 7-1 we see that the corporation as inventor began to assume dominance in 1934. It wasn't until the mid-to-late 1950s that patents issued to corporations began to far-outnumber those issued to individuals. This is a trend which continues today as companies and the government continue to invest heavily in formal research and development. For example, Allan Hall (1985) tells us that since the 1970s research and development spending within organizations has been on the increase. And, Jeffrey Mervis (1994) instructs us that Clinton proposes increased spending in research and development as a way to improve technology and boost economic growth. What is not to be denied however is the power that patent ownership gave to corporations at the expense of the individual.

CONTRACT LAW

The stormy relationship between Western Union and the Bell Telephone system did not end with the patent infringement suits. Western Union was also largely concerned with keeping the Bell Company out of the telegraph field, but these two technologies were intimately intertwined. Western Union owned shares of stock in certain Bell companies. Especially important were the shares Western Union had in the New York Telephone Company. Such substantial interest in the hands of a competitor was interfering with the management of the system as a whole. (Western Union opposed the plans of A.T. & T., who controlled all New York Telephone Company stock except that held by Western Union, to reorganize N. Y. Telephone and incorporate it into the A.T. & T. system). In 1909, A.T. & T. had the opportunity to buy 30% of the Western Union stock, which it did, hoping to open the door for some joint ventures between Western Union and A.T. & T. The Bell system became the stockholder of record on Western Union stock books on November 16, 1909. At this time, a contract was drawn up (by A.T. & T.) and signed by Western Union, was intended to be a constitution of sorts, governing the operating relationship between the two companies. The idea behind this agreement was to pool resources and facilities in order to eliminate duplication and waste, and thereby providing greater profitability and development opportunities for both companies. However, this pooling of resources led to major problems.

Figure 7-1. Number of Patents Issued by patentee 1901-1970

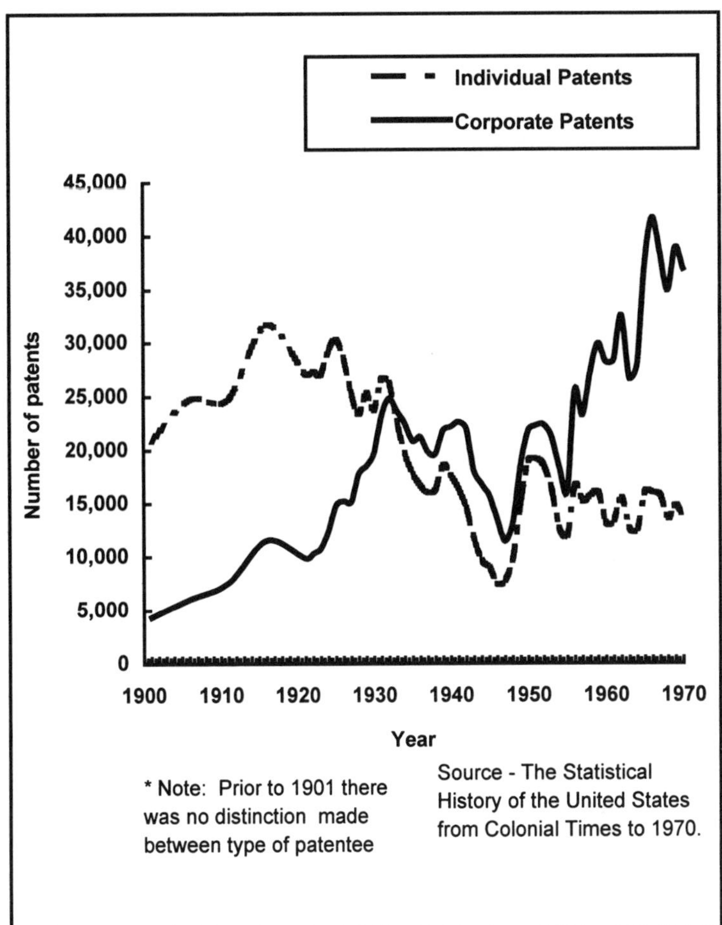

when Western Union was "divorced" from the Bell system in 1913. For example, the engineers of Western Union and the Bell system (which included A.T. & T. and Western Electric) pooled their experience and jointly studied the telegraph printer, and many patents were issued. The rights to use these inventions were granted to the Bell system after the divorce, which effectively meant that Western Union's competitors (including the telephone company itself) could manufacture and sell printers jointly developed. Western Union found itself stripped of its minority interest in the New York Telephone Company and, Western Union lost its premier position in the printing telegraph patent field. This was a blow economically to Western Union because, during its brief contract period with the Bell system and the agreement to share facilities, Western Union furnished the Bell system with much of the background knowledge that it has since used in developing long distance business

A few words need to be said about the rapid divorce between A.T.&T. and Western Union, as it appears that the relationship benefited A.T.&T. the most. Despite the extensive preliminary legal investigations made by A.T.&T. in 1909 which concluded that there was nothing in the contract with Western Union that violated the Sherman Act, the telephone company decided to divest itself of Western Union because other independent phone companies were complaining about A.T.&T.'s ability to unfairly dominant the telecommunications market due to its relationship with Western Union. On January 13,1913 the Interstate Commerce Commission ordered an inquiry into the history, management, operations, rates, rules, regulations, and practices of the telephone and telegraph companies. The A.T.&T's "Supplementary Report . . ., 11" states that

> " . . . while its operating contracts with Western Union was lawful, as counsel said they were, perhaps its control [A.T.&T. shared a president and several board of directors members with Western Union] over Western Union (a different question) was clouded with legal doubts and in such a situation the Telephone Company was unwilling to proceed under the contracts. Having achieved one of its major objectives, namely recovery of the Bell shares, it decided to sever the special relationship with Western Union."

We saw in chapter 2 how important inter-industry linkages were using contracts between Western Union and the railroads as an example. It showed how such contracts facilitated the emergence of Western Union as the dominant telegraph company. This section is meant to illustrate the importance if intra-industry contracts in the struggle for market dominance in a given industry. In this case, however, Western Union failed to maintain its dominant position in the communications industry.

THE TELEGRAPH AND THE LEGAL SYSTEM

The fact that telegraphy was a communication technology of record meant that it could be called into use in legal matters. Its use in a trial, infrequent as this occurrence was, clearly exemplifies how novel uses were continually invented; we see users as innovators in a sense. The following example will elucidate this relationship. *The Journal of the Telegraph* (February 1, 1869) contained an article written by Judge Caton on a trial by telegraph. It happened that the Honorable J. D. Caton, who was in Ottawa, Ill., allowed a man, who was in Chicago and accusing another man of larceny, to send an affidavit via telegram. Judge Caton then used this affidavit to issue a warrant to arrest the accused man who was now in Ottawa, Illinois preparing to board a steamer to California.

Upon questioning the man, the Judge was not convinced of his guilt. To send the man back to Chicago to stand trial seemed to have the possibility of doing the accused a grave injustice as it would disrupt his travel plans with this friends; a delay they could not afford. Judge Caton had an idea: "Soon it occurred to me that if I could arrest a man by telegraph, I might try him by telegraph"—which he immediately did. Testimony was given via the telegraph, with the written telegram acceptable as legal testimony. Within two hours, the case was settled, the defendant was acquitted and free to board his steamer. Judge Caton also wrote "This was the first, if not the only trial, I had ever heard of being conducted by telegraph eighty miles off . . . I am one of those who believe that the law is capable of adapting itself to all the improvements of advancing civilization."

It is clear that, though this was not a use dreamed of by the original inventor of the telegraph, it is a good example of how people manipulated technology to their advantage, that is, how human agency influences the development of technology. It further illustrates how a technology, in the process of embedding itself in the culture, finds continuing and imaginative applications of that technology. Some people were even aware of other uses to which the telegraph could be put regarding the legal system. For example, a *Harpers New Monthly Magazine* (1873: 358) article states "Almost every new season brings forward some new application of telegraph. A complete network or wires now connects the stock exchange with the brokers' offices and the leading hotels, and the business done at the board is printed off in more than a thousand places in New York simultaneously. There is no reason why it should not be employed in a similar way to communicate to the offices of lawyers the progress and proceedings and the call of the calendar in the courts of law." Such a use would greatly enhance the lawyer's ability to manage his time, hence allowing for greater profitability. Though a telegraph system was not set up between the court house and lawyers offices, the Law Telegraph Company was established in 1874 in New York City as a telegraph service for intercommunication among law firms.

One final note concerning the relationship between the courts and telegraph technology. Being a communication technology of record, it was almost inevitable that the telegram should come to be considered evidence. This became problematic. Usually, an original document is considered proper as evidence. This was not an easy determination to make where telegraphic dispatches were concerned. Which was the original—the message a person wrote and handed in at the telegraph office, or the one the receiving operator wrote out and subsequently was delivered? The two messages could differ due to mistakes made in transmission. In his work *A Treatise Upon the Law of Telegraphs*, Scott (1868: 296-297) wrote "It is becoming more and more important that the rules governing negotiations made by telegraph should be clearly defined and settled, as contracts thus made are increasing in magnitude . . . [Telegraph lines]are now largely employed, not only in transmitting friendly information and general commercial news, but messages which are the only evidence of the negotiations of contracting parties in their business transactions." *The Operator* (January 15, 1879), in a report on this issue, states: "To apply the old-fashioned rules of evidence to telegraphic dispatches is quite impractical without some modifications, and the courts are not yet through with the task of determining what modifications shall be allowed." However, the courts seemed to be leaning towards favoring the rationality of business over the individual. This same article further states "A great variety of questions arise in respect to whether messages were, in fact, sent and delivered as they appear to have been, or must every little fact about the transmission be proved by calling witness. And, the courts are, upon the whole, coming to the rule that they will act in part on the probability that telegraphic business is conducted with regularity, accuracy and success, and will call upon whoever disputes the transmission of a message which appears all right to prove his objections." The rationalization of the telegraph business via the use of uniform rules governing the transmission and delivery of messages, as well as the use of standardized forms, most likely influenced the direction the courts was leaning towards regarding the acceptability of telegraph transmissions as evidence.

The patent fights of Morse and those between Western Union and the Bell Telephone Company were among the first of their kind. They were the first patent battles fought within the context of the newly emerging rational organization, which also had an interest in rationalizing the innovation process. Institutionalizing research and development, that is placing it within the bureaucracy of the firm, allowed for greater control. It ensured patent ownership and was meant as a way of avoiding heroic claims to invention. These initial battles formed the backdrop against which the corporation evolved as "inventor" (as Noble, 1977, labels it). It was a lengthy evolution process, but by the mid-1950s, we saw a strong trend toward rationalized research, that is research taking place within the firm and leading to the granting of patents.

With the rise of bureaucratically organized firms, it was almost inevitable that relationships among firms would also be rationalized. The formal contracts between the firms reflected the struggles for market dominance. In the case of the telegraph, the contract Western Union made with the Bell Company ultimately undermined Western Union's position in the communications industry.

The section on the telegraph and the legal system serves to illustrate the way telegraph technology touched even the minutiae of social relations. None of the examples given radically altered the way law was practiced, but each case did affect the relations among those involved due to telegraphic technology. Perhaps this is one of the hallmarks of a thoroughly diffused technology, that once the dramatic impacts are acknowledged, the technology can still affect the relatively less significant day-to-day relationships.

VIII
The Telegraph, Nationalism, and the Global Community

Much attention has currently been paid to the concepts of the global community and the global economy. In order for any nation to become a participant in the global community, it had first to develop a strong national identity. Telegraphic technology aided in further developing this spirit of nationalism which had its roots in the American Revolution in 1776 and the War of 1812, and which was especially crucial to America in its infancy.

With the laying of the Trans-Atlantic cable the world shrank. No longer were the far reaches of the world alien places. Telegraphic communication was ideally suited for encouraging international communication and creating a greater awareness of current events. This chapter will look at the role the telegraph played in the evolution of nationalism (as it relates to promoting federalism over state identity) in America, as well as the ways it contributed to the creation of the notion of a global economy and the global community. It also examines other innovations in communication technology as part of the ongoing evolution of the global community.

AMERICAN NATIONALISM

The telegraph is credited with breaking down the local, individual character of American communities and thereby promoting wider social cohesion of American Society. As one observer put it in 1846, "The power of the States will be broken up in some degree by this intensity and rapidity of communication, and the Union will be solidified at the expense of the State sovereignties. We shall become more and more one people, thinking more alike, acting more alike, and having one impulse." "Lightening Spirit of the Age" (1853: 24-25) In other words, it was perceived to have an homogenizing influence, emphasizing the similarities and commonalties among the diverse people of American society. It was felt that the new facility of communication helped destroy the old sense of community and self-government. The autonomy of the towns and cities was jeopardized by a greater capacity for centralized control made possible by the revolution in communication. A *National Intelligencier* editorial of February 5, 1847 pointed out that there had been some agitation within the country about moving the seat of government closer to the geographic center of the country, for the people of the time thought centrality was essential to good government. This editorial makes the point that "with the invention of the telegraph, centrality is no longer an issue. Washington, D.C. is as close to San Francisco and Oregon as it is to Baltimore or New York." Politically, the telegraph made it possible to administer as a federal unit, the vast territory that sprawled from the Atlantic to the Pacific. The telegraph, with its continuous more rapid transmission of information made for greater public awareness and involvement in political affairs and aided the emerging sense of nationalism. This trend toward "national" thinking can be illustrated by events within the telegraph industry itself. Gabler (1988: 146) tells us that, in the midst of the Civil War, in November, 1863, telegraph operators met in New York City and formed The National Telegraphic Union.

> "National is a key word here, for the war (and subsequent Reconstruction) had a marked centralizing influence on the country. Three national labor unions had emerged in the 1850's, but thirty one appeared during the

> 1860's and 1870's. Sectional conflict, the reintegration of a chastised South into the Union, and an increasingly powerful and activist federal government did much to make labor leaders think nationally. So did the shifting economic emphasis from local to national markets. And no industry better represented this crucial change than the telegraph, no firm better than the Western Union."

Additionally, events that previously were often history by the time news of them reached distant places, became subject to instant political opinion and intervention. As Kirk (1892) states "To the arguments of those who anticipate separation on account of distance and extent of territory, the telegraph replies by diminishing space and time in such a way that . . . Electric wires will bring the thoughts of the most distant states together in a few hours . . . " The telegraph helped to facilitate the process of creating a spirit of nationalism by opening up a new avenue of communication which made people feel closer to one another. This was especially important to the pioneers who lived on the frontier and were helping to establish the limits of our national boundaries. Clearly demarcated boundaries were an essential element in establishing the United States as an important member in the world polity.

Another social effect of the telegraph that aided in the evolution of a spirit of nationalism was its contribution to the spirit of democracy on a cultural level. The semaphore (or optical telegraph) had grown up as a tool of big business. The first optical telegraphs were set up along the harbors on the northeast coast to signal ships approaching the harbors. It had been a device with which to get closely guarded and valuable shipping news to merchants as quickly as possible so as to act to their advantage before other had the reports. The telegraph broke that tradition, making such information available to the mass of people. Between cities the horse-driven vehicle and the courier were agencies of the wealthy and the aristocracy. The telegraph, like the railroads, became a "common carrier", serving rich and poor alike. The railroads and the telegraph, together, made the population mobile enough to expands their horizons through affordable travel and intercourse with remote areas of the country. In a sense, the telegraph and railroad allowed for some what of a leveling off of the class system, fueling

American belief in democracy and equal opportunity for all the people of the country. Additionally, it could be hypothesized that this added to the myth-making process of conceiving of American society as classless or, of one class, the "Imperial Middle" as DeMott labels it.

The next innovation in communication technology also added to the emerging sense of community. Though the telephone only diffused slowly as a tool of business, it diffused rapidly as the sociability aspect became recognized by the telephone companies and they started marketing aimed at this use. Initially, the telephone company promoted the telephone to potential residential customers by telling them that it could be used for such things as: ordering dinner, hastening a delivery, inviting guests, reserving tickets, calling a carriage, making appointments, home shopping, calling a physician. These practical uses were not of a kind that engendered a sense of need in the public's mind. All these uses could also be satisfied via the telegraph, which had the added advantage of only charging when the service was used (as opposed to having to become a subscriber). Also, the telegraph did not require an investment in equipment. Fischer (1992) tells us that it was the telephone as a social device that was mainly responsible for its widespread diffusion. He tells us that there was a general sense among the American public that the pace of life was quickening, and that there wasn't enough time for socializing, especially with those at a distance. Eventually, the telephone company picked up on this sentiment and started a marketing campaign in the 1920s that stressed sociability over practicality. This was the use that eventually got residential customers to subscribe. This serves as a good example of the dynamic relationship between technology and various aspects of social life. The controllers of telephone technology wanted to push the technology toward practical uses. Owing to the fact that many of these men had also been prominent in telegraph technology, the advocated some of the same uses because these helped drive telegraph diffusion. However, consumers were not persuaded by this argument; the practicality aspect of this new technology was not good enough to create a sense of need that would then fuel diffusion. In one sense then, it can be said that resistance on the part of consumers was able to pull the technology in another direction, which eventually led to widespread diffusion when the gatekeepers recognized and responded to this need.

A similar trend can be seen in regard to personal computers in the home. Initially the PC was marketed in terms of its practical uses, as a powerful workstation and as a tool for more efficient home management (one could use it to balance a checkbook, computerize recipes, keep a calendar). However, none of these uses really captured the imagination of the mass of consumers to the point where they felt justified in making a large investment in such equipment. It was not offering the average consumer the ability to do something that he/she couldn't do in any other way The use of the PC as a new form of communication (that is, in terms of the ability to hook into the data superhighway) in combination with lower costs, has greatly aided in the diffusion of personal computer technology.

THE TRANS-ATLANTIC CABLE, THE GLOBAL ECONOMY AND THE GLOBAL COMMUNITY

Cyrus Field, most famous for his part in getting the Atlantic cable laid, was hailed by James Bright *The Echo* (July 13, 1892) as "the Columbus of modern times, who, by his cable, had moved the New World alongside the Old." This description adequately expresses the sentiment of the time toward the connection of Europe and America via technology. A *London Times* article (July 14, 1892) discussing the successful laying of the transatlantic cable in 1866, states "It has transformed the whole method of dealing with international questions; it has transformed the whole conditions of trade; it has, through newspapers which print the daily outcome of it, changed the whole attitude and mind in which Europe and America stand to one another. Thanks to the cable, New York and Chicago seem as near to us as Brussels and Berlin."

Global Economy

DuBoff (1980: 460) writes that there was a growing demand in the United States for improved business services between 1844-1860. He informs us that technology answered this demand through the telegraph" forging extralocal and interregional links among merchants, bankers, brokers and shippers." One important step DuBoff informs us, in increasing information about market conditions was the birth of intercontinental, that, overseas, telegraphy in 1858. (It would be another 8 years before a sustained cable system was operating due to technical difficulties with the actual laying of the cable along the ocean floor.). Prior to telegraphic communication on an international level, merchants had to wait for ships returning from Europe for the cash for cargoes already shipped or for goods they were importing. Knowledge of the world market was distorted by the amount of time between the shipping and receiving of goods, and the lack of up-to-date information of world affairs. The telegraph made it possible to cable money from point of import to point of export. It reduced the risk of bad or unhonored credit, and, combined with the development of steamships, made it possible for companies to efficiently coordinate and direct the purchase, transport, routing, insuring, receipt and distribution of goods. Merchants were able to gain more control over their business—respond to supply and demand and keep an eye on events taking place in foreign markets that could influence their business. With swift communication, merchants could plan and anticipate the movement of their goods; the precise scheduling of services made multiport voyages possible.[2] Within the United States, the telegraph also increased the availability of price information. Traders formed eleven commodity exchanges in the larger cities between 1845 and 1871. These exchanges used the telegraph to buy and sell commodities while they were still being grown or transported. This directly contributed to reducing uncertainty and reducing the costs of movement of American crops. (*Journal of Urban History*, 1987)

The telegraph also facilitated the centralization of America's securities markets in New York City between 1850 and 1880. By 1860, Wall Street exchanges were connected with and set prices for, every major city in the United States. The telegraph was used to negotiate back and forth throughout the day, thus accelerating the rate of capital turnover and reinvestment. Additionally, the Gold and Stock Telegraph Company was established in 1867 (which Western Union bought out by 1871) and gold indicators directly reported price information from the gold exchange to the offices of subscribing brokers eliminating the need for messengers to run back and forth.

Additionally, the telegraph, along with railroads and steamships, opened up previously inaccessible areas of the globe to trade. For example, Johnston (1882: 167) tells us that the exploration and illumination of Africa, the once Dark Continent, was made possible by the collaboration of these three inventions. "Nothing has been more remarkable in the history of the last few years than the progress of discovery in the continent of Africa, which promises to shortly open it up fully to the operations of trade, aided by the steamboat, the locomotive, and the electric telegraph." Though the actual impact of telegraphy on the farthest colonies may have been minimal, and certainly, economically, connections with Europe were much more important, we see reflected in the preceding quote the optimism and hope vested in technology innovation which in itself, then becomes the impetus for further research and invention.

In September of 1905, *Scientific American* reported that since the opening of the international telegraph, that is, since the laying of the transatlantic cable, imports and exports had grown from $220,000 to $1,600,00. And, that within the United States, commerce had grown from $1,000,000,000 to $25,000,000,000. Surely, other factors contributed to the spectacular increase in commerce, but the editor of the *Scientific American* attributed it largely to the availability of the telegraphy.

The invention and introduction of transatlantic telegraph put an end to the local market structure which had, until that point, been kept separate by lack of prior information about supply and demand. Telegraphy, the railroads, and the steamship, three complimentary technologies, paved the way for the creation of the global economy and the recognition of America as a commercial entity in the world market.

The Global Community

As noted, it was felt by some that the instantaneous transmission of news brought the country a little closer together and gave the citizens more of a sense of national identity. The telegraph was also believed to have collapsed the world community. James D. Reid (1853) with the spread of telegraphy in mind, "In all these enterprises we see the means of a greater development of human progress, the enlightenment of society, the elevating of the public mine . . . With intelligence comes freedom. With freedom comes sympathy with interests in the necessities of society. Man becomes less a citizen than a brother." Electronic communication made it possible for remote areas of the world to be apprised of events no matter where they occurred. People no longer had to wait for the arrival of ships to find out what was happening abroad. Instantaneous news became the journalistic standard. For example, Americans were able to get continuous reports about the course of the Franco-Prussian War in 1870 because the *New York Tribune* had a team of reporters, covering the war from the battlefields (a technique first developed in the Civil War), whose reports were sped home by telegraph. The telegraph decreased the isolation as well as the privacy of countries. Intercontinental intercommunications created strong interdependencies and vulnerabilities. Both friend and foe alike were able to gather more information, and do it more rapidly and currently. The more current information one has on another, the more vulnerable the other party becomes. Additionally, the mass of people were finally getting a global view. The telegraph, via news reporting, allowed people to get a more culturally realistic image of life in distant places.

President Buchanan's congratulatory message on the laying of the Atlantic cable of 1858 stated in part "may the Atlantic Telegraph under the blessing of heaven, prove to be a bond of perpetual peace and friendship between the kindred nations, and an instrument, destined by Divine Providence, to diffuse religion, liberty, and law throughout the world." and Queen Victoria's message stated that it is "the ardent wish that peace and good will should reign between hitherto unfriendly nations, born to the same blood, speaking the same tongue, and rejoicing in the same faith." *News* (London) (July 22, 1892) That these wishes for world peace never come to fruition, leads us to raise a few questions concerning electronic communication. After giving the telegraph due credit for all its social benefits, it is a matter of common observation that it has not brought about the millennium. Innovations in technology still have their limits. Though we look to technology in our quest for increasing rationalization of the world, understandings and misunderstandings among nations are still fostered by people: merchants, artists, diplomats, educators, students, tourists, writers. The good that they accomplish, as well as the damage that they do still fosters certain beliefs and attitudes among people. Books (in as much as they contain "the wisdom of the ages") and people, were still important agencies of international communication that the introduction of the telegraph and for that matter, other subsequent communication technologies, did not negate. But we are still trying, and maybe that is a driving force behind innovation.

The telegraph paved the way for international communication, a first link in the creation of the global community. However, the telegraph, and later, the telephone, faced one major limitation in making international contact truly available to the general public: they were forms of point to point communication which required prior knowledge. One had to address a telegram to a particularly party in a particular place; one had to call a specific phone number. Consequently, though global communication was a technical reality it was in actuality, only so for those who already knew someone in another country or somehow had prior contact with another culture. The Internet[3] is the first innovation in communication technology that has the potential to make global communication a reality on the level of the general public. The reason this becomes feasible is because Internet (and similar services) offers anonymous point to point communication. This means that with the introduction of Internet service into homes via the personal computer, it is now possible for anyone to communicate with anyone else, anywhere in the world via a bulletin board. This means that you do not need to have a specific point of contact in order to use the service. Specific point of origin to non-specific point of destination is what has made the idea of a global community more than just a possibility.

Much research has yet to be done on the kind of community being created by the Internet and related services, but clearly the diffusion of personal computers into the home is an important component of this research. One would want to know who is using the Internet, that is, the demographic characteristics of the new members. (We already know that the initial users were the government and scientists and academics). Is it a classed based phenomenon? Is the communication potential of the personal computer one of the things that is causing a new diffusion of this technology? In other words, what can be said about the diffusion of this technology in terms of what it offers—does this communication potential of the PC make it more marketable because it can now offer something that people cannot do in any other or more convenient way? Remember that some of the home management functions that the PC made possible did not sell as well as expected. For example, people did not consider it worth the investment to use the PC to do such things as manage the household checking (glorified calculator), or as a repository for recipes (a total impractical use unless the PC happens to be in the kitchen, and even then, the need to keep foreign materials out of the unit itself makes this impractical as well.) Also, there appeared to be no savings in having to print out a recipe before using it. The question is, does Internet use represent a new use for the PC that would substantially aid in its diffusion? ? Recent scholarship seems to suggest that this is the case. The Internet has been around since 1969 when the Department of Defense started ARPNET. During the 1970s and 1980s the Internet served university based research that was primarily government sponsored. In the mid 1980s the National Science Foundation (NSF) incorporated the Internet into its research centers According to Juliussen (1996) there were only about 1,ooo computers connected to the Internet in 19855. Widespread diffusion of the Internet was tied to widespread diffusion of the personal computer. Internet access via the PC offered people a use for the PC that they could not achieve in any similar way. 1991 saw the for-profit, commercial Internet connections become available to the general public and 1993 saw the appearance of a standardized software program that interfaced with the Internet's World Wide Web and simplified Internet access. People responded. Juliussen and Juliussen (1996: 505) write that "The PC market has had exceptional growth in the last two years primarily caused by strong home/consumer sales." The Computer Association stated in 1995 that "The personal computer has finally come into its own as an accepted and desired member of the American household. 1995 was the first year in which dollar sales of PCs were higher than that of all television sets..." (Juliussen and Juliussen (1996: 507) As the model of innovation and diffusion developed here suggests, the widespread diffusion of these technologies depended on standardization and incremental improvements in the technology.

The telegraph was a significant factor in practically and psychologically uniting the vast area that comprised the United States. It promoted the idea of a federal identity over a state-centered identity by keeping people across the nation up to date with events going on in all areas of the country. The telegraph, in combination with the railroads, made information and travel more accessible to the masses, engendering a sense of common identity. The social cohesion fostered by telegraph communication was further enhanced by the next innovation in interpersonal communication technology, the telephone.

The extension of telegraphic capability internationally fostered globalization by strengthening and extending economic ties and by giving business greater control over shipping and markets. It put an end to the local market structure.

Just as the telegraph fostered social cohesion on a domestic level, it enhanced the feeling of community internationally. Instantaneous international news reporting was especially responsible for bringing the world a little closer together. The telegraph was the catalyst for international communication that continued with the telephone, and continues today with the Internet.

NOTES

1. Vail started out as a telegraph operator and his uncle was Alfred Vail, nemesis of Morse during his patent battles.,

2. Field (1992) discusses the impact of the ability to plan and precisely control the movement of goods on domestic industry. Field tells us that not all industries were able to benefit significantly from the telegraph, but among industries where timing was crucial, the benefit was great (e.g. the chilled fruits and vegetables and meat processing).

3. It is interesting to note that we, as a society cannot escape the tendency to subscribe to the heroic theory of invention. A recent example is the article by Katie Hafner (*New York* Times, September 25, 1994) entitled "For Father of the Internet, . . . " Though the article does mention that there have been many people responsible for the development of the Internet it emphasizes the role of Vinton Cerf: ". . . but it [Internet] would not have occurred without Vint."

Conclusion

Technology is the tool with which we manipulate our environment in our perpetual quest to gain greater control over life. Communication is essential for the process of control. Manufacturing technologies and consumer technologies have received much attention in the literature on technology diffusion. It was precisely because there was little diffusion research done on communication technology that I chose it as the subject of this work. In so doing, it seemed only appropriate to begin by examining the first modern communication technology which seriously altered the ability to control and mediate experiences with other social institutions.

Technology innovation is a process punctuated by different phases: origin (identification of needs/problems, research), diffusion (development, commercialization, adoption), and impact (consequences). Though these phases often tend to be treated as if they were discrete entities, they are in fact continuous and reciprocal. Consequently, the model for technology innovation presented here is an adaptation of Rogers (1983) in which specific aspects of various phases were modified. It was necessary to distinguish between general needs and specific needs as each of these will influence the diffusion process in terms of the potential adopters. Invention as the result of general need is hypothesized to have an amorphous pool of potential adopters. Invention as the result of specific need is hypothesized to have a ready-made pool of potential adopters. These factors will ultimately impact on the shape of the S-curve (slow/fast initial diffusion). Also, the invention and development phase was modified to distinguish between radical, macro inventions and incremental, micro inventions as these too will affect the rate of diffusion in terms of the social and cultural barriers to diffusion. And, finally, the diffusion phase encapsulates under six headings the variables that different disciplines have espoused as important to the diffusion and innovation process.

Though various disciplines have focused on the diffusion phase of the innovation process and offer explanations for the reasons for the rate of diffusion appearing as an S-curve, none of these seemed particularly well-suited to interpersonal communication technology. Consequently, a model of the diffusion phase that is specific to such communication technology was presented. The hope is, that this model will advance the general understanding of the innovation process from a more detailed understanding of the specific applications. This model posits five stages of the diffusion process for an interpersonal communication technology: ownership, development, infrastructure building, expansion, and decline. The history of the trajectory of the telegraph aptly illustrates the stages in this model, and suggests that the telegraph was prototypical for other interpersonal communication technologies. By identifying the tasks/issues to be resolved at each stage, this model helps to predict the general pattern for interpersonal communication technologies.

This model also suggests that in order to understand the diffusion of an innovation, one must look at the pattern of conditions that preceded the actual diffusion process, because these variables will inform the pattern of variables in the diffusion process. For example, the telegraph was a radical innovation developed in response to general need. Consequently, there was no guaranteed pool of potential adopters. This meant that a tangible need had to be illuminated or evoked. In creating or uncovering this sense of need, cultural norms and values, as well as political and economic ideology were important interactive variables in initiating the diffusion process. Consequently, it is hypothesized that any innovation in interpersonal communication technology which matches this initial pattern will have a slow beginning to its S-curve when modeling diffusion rate. As shown in Figures 2-4 and 2-5 both telegraph and telephone technology have a slow beginning to their S-curves.

It is also important to point out that the diffusion process is also greatly influenced by whether or not the innovation requires the building of a new industry. This adds another dimension to the diffusion process, one in which economic considerations and constraints (especially profitability) play a major role. Telegraphy had a slow start because it was an innovation which had systemic needs and a structure in which to diffuse such a technology needed itself to be innovated. The existence of competing firms, the lack of standardization, and the lack of a system all colluded to retard the initial diffusion of the telegraph. Thus we see the rise of large scale bureaucratic organization and the evolution of the business monopoly (with their attendant characteristics), as integral to successful telegraph diffusion. A similar pattern was observed for telephone technology, reinforcing the idea that this model is applicable to any radical interpersonal communication technology.

The diffusion model of interpersonal communication technology being proposed in this work incorporates the reciprocity among stages into the diffusion process. The invention that initially becomes commercialized is not necessarily the invention that gets diffused. Continuing improvements affect the diffusion process. Using telegraphy as an example, the telegraph that was in use in the 1879s was able to transmit more messages per telegraph line than that introduced in the 1840s. This meant that more messages could be handled in shorter periods of time, and this was crucial when marketing an innovation whose selling point was the mediation of time . And, as a further example, one can look at telephone technology. If one looks at the diffusion of telephone technology, the telephone introduced by Bell has only a vague resemblance to the telephone of today, both in appearance and capability. In its nascent phase, there was no choice for subscribers, When you elected to subscribe you got basic telephone calling service and were leased a black phone. Today, telephone subscription comes with a menu of options, from choice as to owning or leasing equipment to (services you want to receive, call waiting, call forwarding, call answering, touch-tone, etc.).

The telegraph, as a case study, was also used to illustrate the impact of a new technology on American society. It was found that the more diffused throughout a culture a communication technology becomes, the more it becomes embedded in the various institutions within society and the more novel its applications become. Hence, one could say that even as a technology continues to diffuse, it impacts on itself as an antecedent technology in a sense. Again, this points to the reciprocity among variables that can affect the diffusion process. Additionally, consequences of an innovation become part of the accumulative culture of society and can act as fuel for continuing research and development. It is this continuous flow of interrelationships among the various stages as well as the different variables within these stages that the modified model of the innovation process I offer is meant to depict.

The impact of an innovation can vary from one limited in scope to one, which, over time, can greatly affect the fabric of the social structure. It is believed that any successfully diffused innovation increases the rationalization of the world, bringing it more and more under human control. Each innovation will alter relations of control, sometimes on a limited basis and sometimes on a societal basis. If one looks retrospectively at the impact of telegraphy, there appears to be almost no institution that was not affected by this invention. Telegraphy helped to change norms and define culturally appropriate relationships with time and instantaneous communication. Later outgrowths of the invention of telegraphy was the fax machine and electronic mail. Telegraphy affected the structure and organization of work, serving as the prototype of the modern business firm and introducing new jobs into the social structure. Telegraphy has been credited with beginning the field of electrical engineering within the business community, and being the first conglomerate. Telegraphy affected the social stratification system by contributing to the emergence of the lower middle class (via the new types of jobs), and in terms of defining women's roles and place in the world of modern work. From a control perspective, technology is a double-edged sword. As someone or some group is gaining control and power, someone else or some other group is losing it. Consequently, technology can be used to reinforce the existing stratification system in society. Finally, telegraphy aided in the centralization of control within society. Examining the impact of technological innovation is an integral part of studying the innovation process. The changes in social structure that inventions engender (whether subtle or dramatic) influence the history of succeeding innovations. The Internet seems to be the communication innovation which most parallels the telegraph in terms of its impact on society. One only has t peruse the current literature to see the breath of social experiences encompassed by the Internet. Literature exits on such topics as health and the Internet, Internet advertising, cybermarketing, sports on the net, education and distance learning on the Net, and the list goes on.

One can get a sense of how deeply embedded in society the PC and the Internet have become if one compares the topics discussed in Part II of this book with information available today. For example, a brief comparison with the six topics discussed here yields the following: Formal organizations- telecomputing has affected the structure of the modern office in that many workers need not be physically present in the office in order to perform their jobs. They can work from home/car/train/ plane. Aside from affecting the structure of the office, it blurs the line between work and leisure time and begs the questions of promotion and evaluation of work within the workplace. The political system. Economics and culture - an example that demonstrates that the internet is having an impact on these issues is the appearance of books dealing with these topics. Some examples are : Neil Barrett's (1996) book *The State of Cybernation: cultural, political and economic implications of the Internet*, *Netspeak: the Internet Dictionary* (Fahey, 1994) and *Backslash* (William Lovejoy, 1996) which deals with adventure and detective stories based on the Internet and computer network fiction, and The Art of Computer Conversation by Gaines nad Shaw (1984) in which they discuss computer mediated conversation. These are but to name oa few of the may books on these topics. Additionally, one only has to sign onto the Internet to see the proliferation of business and advertising on the Net. This information superhighway is adding a dramatic new dimension to the marketplace (analogous to the changes in the marketplace facilitated by the telegraph.) Judicial system - Benjamin Wright's book *The Law of Electronic Commerce: EDI, E-Mail, and Internet technology, Proof and Liability* includes a discussion of evidence law, liability law, and contracts as these relate to the titled technologies.

Surely, the Internet is embedding itself in society and we are living in the midst of the changes in control and social relations that it is engendering. I would dare to posit that the Internet is changing society on a scale comparable to that of the telegraph, and like the telegraph, is only the beginning of many incremental changes in technology which will profoundly alter our society. Time and research will assess the extent and impact of that change.

Bibliography

Aronson, Sidney 1977. "The Lancet on the Telephone 1876-1975" *Medical History* .21, No. 1, (January):69-87.

———. 1971. "The Sociology of the Telephone" *International Journal of Comparative Sociology*. 12 (September): 153-167.

———. Personal, unpublished notes on the telephone and telegraph.

Axline, Andrew and James Stegenga 1972. *The Global Community*. New York: Dodd, Mead & Co.

Bellah, Robert N. 1985. *Habits of the Heart*. Berkeley: University of California Press.

Beniger, James R. 1986. *The Control Revolution*. Cambridge, Massachusetts: Harvard University Press.

Berger and Berger 1976. *Sociology: A Biographical Approach*. New York: Basic Books.

Betz, Frederick 1993. *Strategic Technology Management*. New York: McGraw-Hill, Inc.

Blau, Peter and Marshall Meyer 1987. *Bureaucracy in Modern Society*. New York: Random House.

Brady, Jasper Ewing 1899. *Tales of the Telegraph*. New York: Doubleday and Mc Clure Co.

Brown, Ronald. 1975. *Telephone Facsimile for Business*. Stoke-Sub-Hamdon: Ronald Brown.

Brown, Lawrence 1981. *Innovation Diffusion*. New York: Methuen.

Burke, Janmes. 1978. *Connections*. Boston: Little, Brown and Company.

Carter, C.F. il. 1914. "Within a tick of the News." *Tech. World* 21 (April 14): 262-4.

Channing, William 1852. *Municipal Electric Telegraph*. New Haven: B. L. Hamlem.

Chandler, Alfred 1977. *The Visible Hand.* Cambridge, Massachusettes: Belknap Press.

———. 1984. "Comparative Business History" in *Enterprise and History* D.C. Coleman, Peter Mathias, eds. New York: Cambridge University Press.

Channing, William 1852. *Municipal Electric Telegraph.* New Haven: B. L. Hamlem.

Charpie, Robert A. 1970. "Technological Innovation and the International Economy" in *Technological Innovation and the Economy.* Maurice Goldsmith, ed. New York: Wiley Interscience.

Clampitt, J.W. _____. *Echos From the Rocky Mountains*

Clegg, Stewart R. 1990. *Modern Organizations.* Newbury Park, Cal.: SAGE Publications Inc.

Coates, V.T. et al. 1979. *A Retrospective Technology Assesment: Submarine Telegraphy The Transatlantic Cable of 1866.* San Francisco: San Francisco Press, Inc.

Cockburn, Cynthia 1985. *Machinery of Dominance.* Dover, N.H.: Pluto Press.

———. 1983. Brothers: *Male Dominance and Technological Change.* London: Pluto Press.

Cohn, Samuel 1985. *The Process of Occupational Sex-Typing. Philadelphia: Temple University Press.*

Cortada, James. 1993. *The Computer in the United States: from laboratory to market.* Armonk,New York: M.E. Sharpe.

Costigan, Daniel. 1971. *The Principles and Practice of Facsimile Communication.* New York: Chilton Book Company.

DeMott, Benjamin 1992. *The Imperial Middle.* New Haven: Yale University Press.

Depew, C.M. 1896. "Telegraphing Around the Globe." *Scientific American* 74 (May 30): 347.

Du Boff, Richard 1980. "Business Demand and the Development of the Telegraph in the U.S. 1844-1860" *Business History Review* 54(4): 460.

Duckworth, J.C. 1970. "The Role of Government" in *Technological Innovation and the Economy* Maurice Goldsmith, ed. New York: Wiley Interscience.

Ellul, Jacques. 1967. *The Technological Society.* New York: Alfred A. Knopf.

Emery, Michael and Edwin Emery 1988. *The Press in America.* Englewood Cliffs, N.J.: Prentice Hall.

Escher, F. 1911. "Industrial Securities—the Telephone and Telegraph Group." *Harpers Weekly* 55 (Dec. 9): 22.

Field, Alexander James 1992. "The Magnetic Telegraph, Price and Quantity Data, and the New Management of Control" *The Journal of Economic History* Vol. 52, No. 2 (June 1992): 401-413.

Fischer, Claude 1992. *America Calling*. Berkeley: University of California Press.

Flamm, Kenneth. 1988. *Creating the Computer: government, industry and high technology*. Washington, D.C.: Brookings Institution.

Gabler, Edwin 1988. *The American Telegrapher*. New Brunswick: Rutgers University Press.

Gaines, Brian R. and Shaw, Mildred L.G. 1984. *The Art of Computer Conversation* Englewood Cliffs, NJ: Prentice/Hall International.

Garratt, G.R.M. 1967. "Telegraphy" *A History of Technology: the Industrial Revolution 1750-1850*. Charles Singer,ed. Oxford: Clarendon Press.

Giddens, Anthony 1983. *A Contemporary Critique of American Materialism.* Berkeley: University of California Press. 1991. *Modernity and Self-Identity*. Stanford: Stanford University Press.

Gold, Bela 1980. *Evaluating Technological Innovations*. Lexington, Mass.: D.C. Heath & Co.

———. 1977. *Research, Technological Change, and Economic Analysis*. Leington, Mass.: Lexington Books.

Goldman, J.E. 1970. "Setting the Scene One: the United States" in *Technological Innovation and the Economy* Maurice Goldsmith, ed. New York: Wiley Interscience.

Goldsmith, Maurice, ed 1970. *Technological Innovation and the Economy*. New York: Wiley-Interscience.

Gordon, George N. 1977. *The Communications Revolution:a History of Mass Media in the United States*. New York: Hastings House.

Hall, Alan 1985. "Research and Development Scoreboard". *Business Week*. Special Issue (Mar. 22) 164-165

Harlow, Alvin. 1971. *Old Wires and New Waves*. New York: Arno Press and the New York Times.

Holzmann, Gerard and Pehrson, Bjorn 1995. *The Early History of Data Networks*. Los Alamitos, CA: IEEE Computer Society Press.

Hounshell, David 1988. "Why It Wasn't Ma Gray". *Harvard Business Review*. Vol. 66, No. 4: 152-153.

Huges, Thomas P. 1989. *American Genesis*. New York: Viking.

Johnston, W. J. 1882. *Lightening Flashes and Electric Dashes* New York: W. S. Johnston.

Johnston, W. J. 1882. *Telegraphic Tales and Telegraphic History*. New York: W. S. Johnston.

Juliussen, Karen Petska and Juliussen, Dr. Egil 1996. *8th Annual Computer Industry Almanac*. Austin, TX: The Reference Press.

Kash, Don.E. 1989. *Perpetual Innovation*. New York: Basic Book, Inc.

Keys, C.M. 1910 "Rulers of the Wires." *Worlds Work* 19 (March): 12726-9

Kieve, Jeffrey L. 1973. *The Electric Telegraph: a social and economic history*. Newton Abbot: David and Charles.

Kirk, J.W. 1892. "First News Message by Telegraph." *Scribners Magazine* 11 (May): 652-6

Lardner, Dronysius 1855. *The Electric Telegraph*. London: Walton and Maberly.

———. 1857. *The Electric Telegraph Popularized*. London: Lockwood And Company.

Lienau, F.W. 1917. "Progress of the Telegraph." *Nation* 104 (Mar. 8): 294-5.

Lind, Robert S. 1949. "You Can't Skin a Live Tiger" *American Scholar*. XVII, p.109.

Lindley, Lester 1975. *The Constitution Faces Technology*. New York: Arno Press.

Machlup, Fritz 1952. *The Political Economy of Monopoly*. Baltimore: The Johns Hopkins Press.

Marcus, Alan and Howard Segal 1989. *Technology in America*. New York: Harcourt Brace Jovanovich, Inc.

Markus, M. Lynne 1987. "Toward a "Critical Mass" Theory of Interactive Media" *Communication Research*, Vol. 14 No. 5:491-511

Marshall, Walter P. 1951. *Ezra Cornell 1807-1874* His Contributions to Western Union and to Cornell University. Manuscript of the Newcomen Society of America

Merton, Robert K. 1974. *The Sociology of Science*. Chicago: University of Chicago Press.

Mervis, Jeffrey 1994. Research and Development: Growth in Hard Times". *Science* v263 (Feb. 11) 774-6.

Mokyr, Joel 1990. *The Lever of Riches*. New York: Oxford University Press.

Morse, Edward Lind, ed. 1914. *Samuel F. B. Morse His Letters and Journals*, Volumes I and II, New York: Houghton Mifflin Co.

Nelson, William E. 1982. *The Roots of American Bureaucracy 1830-1900.* Cambridge, Mass.: Harvard University Press.

Noble, David F. 1977. *America by Design.* New York: Oxford University Press.

O'Brien, John Emmett 1910. *Telegraphing in Battle.* Wilkes Barre: The Reader Press.

Oliver,P., Marwell, G. and Teixeira, R. (1985) "A Theory of Critical Mass: I. Interdependence, group heterogeneity, and the production of collective action." *American Journal of Sociology*, 91 (3): 54-63.

Parker, J.E.S. 1974. *The Economics of Innovation.* London: Longman Group Ltd.

Pavitt, Keith 1970. "Performance in Technological Innovation in the Industrially Advanced Countries" in *Technological Innovation and the Economy* Maurice Goldsmith, ed. New York: Wiley Interscience.

Plum, William R. 1882. *The Military Telegraph During the Civil War in the United States*, Vol. I. Chicago: Jansen, Mc Clurg and Co.

Pred, Allan 1966. *Spatial Dynamics of U.S. Urban Industrial Growth, 1800-1914.* Cambridge, Mass.: MIT Press.

Pursell, Carroll W. 1981. *Technology in America.* Cambridge, Mass.: The MIT Press.

Reid, James D. 1879. *The Telegraph in America.* New York: James D. Reid.

Robertson, T.S. 1971. *Innovative Behavior and Communication.* :Holt, Rinehart, and Winston.

Rogers, Everett M. 1983. *Diffusion of Innovation.* New York: The Free Press.

Roy, William G. 1983. "The Unfolding of the Interlocking Directorate Structure in the United States". *American Sociological Review.* 48:2 April: 248-257.

Rupp, Robert O. 1990. *Hullo, Anybody There?* Colorado: The Old Army Press.

Scharlott, Bradford 1989. "Influence of the Telegraph on Wisconsin Newspaper Growth" *Journalism Quarterly.* v66,n3: 710-715.

Schwarzlose, Richard 1989. *The Nation's Newsbrokers Volume 1: The Formative Years: from Pretelegraph to 1865.* Evanstan: Northwestern University Press.

Scott, William and Jarvogen, Milton 1868. *A Treatise Upon the Law of Telegraphs.* Boston: Little, Brown and Company.

Sewall, C.H. il. 1900 "Future of Long Distance Communication." *Harpers Weekly* 40 (Dec. 29) 1262-3.

Shiers, George, ed. 1977. *The Telephone: an historical anthology*. New York: Arno Press.

Singer, Charles and E.J. Holmyard, A.R. Hall, Trevor I. Williams, eds. 1967. *A History of Technology: the Industrial Revolution 1750-1850*, Vol. IV. Oxford: Clarendon Press.

Smith, J.E. 1865. *Smith's Manual of Telegraphy*. New York: L.G. Tillotson and Co.

Taltavall, John B. 1894. *Telegraphers of Today*. New York: Mc Breen, Club Press.

Taylor, Frederick 1985. *The Principles of Scientific Management*. Easton, Pa.: Hive Publishing Company

Tebbel, John 1969. *The Compact History of the American Newspaper*. New York: Hawthorne Books, Inc.

Thompson, Robert L. 1947. *Wiring a Continent*. Princeton,N.J.: Princeton University Press.

Townsend, Carl 1984. *Electronic Mail and Beyond*. Belmont, CA: Wadsworth Electronic Publishing.

Trice, Harrison 1993. *Occupational Subcultures in the Workplace*. Ithaca, N.Y.: ILR Press.

United States Bureau of the Census 1971. *The Statistical History of the United States from Colonial Times to 1970*. New York: Basic Books, Inc.

Vail, S. 1896. "First Telegraph." *Harpers Weekly* 40 (Dec. 26): 1294

Vail, S. 1891. "Early Days of the First Telegraph Line." *New England Magazine* 4 (June): 450-60.

Weiller, L. 1898. "Annihilation of Distance." *Living Age* 219 (Oct. 15): 163-78.

White, E. B. 1939. *Telegrams in 1889 and Since*. Princeton: Princeton University Press.

Zerubavel, Eviatar 1985. *The Seven Day Circle: the history and meaning of the week*. New York: Free Press.

———. 1981. *Hidden Rhythms: schedules and calendars in social life*. Chicago: University of Chicago Press.

1850 O'Reilly Papers, MSS. Dept. Vol. 59.

1853 "The Cuban Telegraph." *The National TelegraphicReview and Operators Companion*. 1 (April): 24.

1853 "Lightning Spirit of the Age." *The NationalTelegraphic Review and Operators Companion*. 1 (April): 24-25.

1853 *The National Telegraphic Review and OperatorsCompanion.* 1 (July).

1868 *Manchester Guardian* March 9.

1873 "The Telegraph". *Harper's Monthly Magazine*, Vol. XLVII, 44.

1879 *The Operator* January 15.

1894 "Telegraph and its Inventors." *Scientific American* 70 (June 23): 390.

1904 "New System of Rapid Telegraphy." *Scientific American* 90 (Mar.26): 249-50.

1905 *Scientific American* September.

1906 Department of Commerce and Labor, Bureau of the Census Special Reports: Telephones and Telegraphs, 1902. Washington: Government Printing Office.

1917 "Wires and the War." *Scientific American* 117 (Sept. 15): 186.

1921 "Telegraph Wires as Weather Prophets." *Scientific American* 124 (Jan. 15): 47.

1922 "Call of the Wire." *Saturday Evening Post* 194 (Mar, 4): 12.

1932 *Telegraph and Telephone Age*, No. 9: 194.

1936 "What the Fire Alarm Means to Cities." *American City* 51 (March): 17.

1944 "Preceedinngs from the I.R.E." Vol. 32, No. 8, August

1987 "The City and the Telegraph: Urban Telecommunications in the Pre-Telephone Era". *Journal of Urban History*. 14, 1 November: 38-80.

1850 U. S. Bureau of Census—Population Reports.

1869 Annual Report to the Stockholders, WUTC.

1870 U. S. Bureau of Census—Census Reports.

1880 U. S. Bureau of Census—Census Reports.

1890 U. S. Bureau of Census—Census Reports.

1900 U. S. Bureau of Census—Census Reports.

1902 U. S. Bureau of Census—Special Reports on Telephones and Telegraphs.

1870 *Government Telegraphs*—Arguments of William Orton, President of The Western Union Telegraph Company on the Bill to Establish Postal Telegraph Lines, delivered before Select Committee of the United States House of Representatives, 1870. New York: Russell's American Steam Printing House.

Brooklyn Daily Eagle March 9, 1910.

London Times July 14,1892

News (London) July 22, 1892.

Journal of the Telegraph Vol.2, No.4 January 15, 1869 New York

New York Express June, 1846.

New York Times 1852, 1863-1874.

New York Tribune 1870.

The Echo July 13,1892

———. (undated) *Something Electrical for Everyone*. Catalog, contained in the Western Union Archive Collection, American Museum of History, Smithsonian Institution.

Western Union Archive Collection, (I and II) American Museum of History, Smithsonian Institution, Washington, D.C. Misc. notes, letters, documents and reports Annual Report to the Stockholders 1861-1890, 1905 Booklet "Social Telegrams" Misc. collection of messenger records Misc. collection of social telegrams Rule Books 1870, 1884, 1900 "Supplementary Report On Outline of A.T.&T. Company Relationship To Telegraph Business" Western Union's Place in Management and Technology Trends in Telecommunications

Index